ALLE ZEIT WACH 1842

Jean U. Thoma

Simulation by Bondgraphs

Introduction to a Graphical Method

Springer-Verlag
Berlin Heidelberg NewYork
London Paris Tokyo Hong Kong 1990

Professor Dr. Jean Ulrich Thoma
Dept. of Mechanical Engineering
University of Waterloo
Ontario, Canada N2L 3G1

Thoma Consulting
CH-6300 Zug, Switzerland

ISBN-13:978-3-642-83924-5 e-ISBN-13:978-3-642-83922-1
DOI: 10.1007/978-3-642-83922-1

Library of Congress Cataloging in Publication Data
Thoma, Jean U. (Jean Ulrich), 1927–
Simulation by bondgraphs : introduction to a graphical method /
Jean U. Thoma.
Includes bibliographical references.
ISBN-13:978-3-642-83924-5(U.S. : alk. paper)
1. Bond graphs. I. Title. II. Title: Simulation by bond graphs.
TA338.B6T47 1900
620'.001'1–dc20

Typesetting: Thomson Press India Ltd., India;

2161/3020-543210 – Printed on acid-free paper

Preface

The idea of this new Bondgraph book appeared during the authors stay at Osaka University in spring 1987 as fellow of the JSPS (Japanese Society for the Promotion of Science). Instead of having my 1975 book translated into japanese I felt it needed to be completely rewritten. The japanese edition is under way as a joint-venture with Prof. N. Suda of Osaka University and should be available soon by Corona Publisher in Tokyo as Nr 9 of their series on Modern Control Engineering.

The basic plan comes from my old book on Bondgraphs (Thoma 1975), meanwhile published also in German and Italian. It has been modified from teaching and consulting experience, particularly from the graduate course "Modern Machine Design by Bondgraphs" at the University of Waterloo, Ontario, Canada. Many of the 20 worked examples come from the industrial simulation of real machines, installations and systems.

Since the early years the PC (Personal Computer) has changed the working environment of the modern engineer and scientist, as with many other professions. Today (1990), the programing power of Bondgraphs produces impressive results on inexpensive personal machines.

Bondgraphs stem from the ingenuity of Henry Paynter at MIT (Cambridge, USA) since the 1960ties. They are a representation of engineering systems on paper by letter elements and their interconnections, called bonds. They posses a myriad of advantages (Margolis 1985) which become impressive, even world shaking with moderately complex simulations. We therefore expect a great increase of bondgraphing in the future.

Our intended readers remain the same: Any person, student or engineer in industry who needs to simulate systems with mechanical, hydraulic or electric parts. Due to its application oriented character, bondgraphing is most useful for increasing the efficiency of new product design. Further, the book can be used for Bondgraph courses and will be by the author for his courses Modebond and Şimbond (Appendix 1).

We expect no prior knowledge of Bondgraphs from our readers, only a feeling for control engineering, including block diagrams, transfer functions and stability. Interest in interdisciplinary engineering and analogies will help.

In the first chapter we describe graphical system models in general whilst Bondgraphs are introduced in Chapter 2. The following 3 chapters treat the writing of Bondgraphs for mechanical, hydraulic and electric systems with worked examples. Computational details and elimination of numeric difficulties come in chapter 6, in order to also teach the practical aspects of simulation.

Chapter 7 speaks about thermodynamics by Bondgraphs with extension to thermal engines and chemical networks. More involved questions are treated in chapter 8 and the final chapter brings further worked examples.

The examples are worked with the TUTSIM (1988) program, which is a fast and powerful program for microcomputers. However, other programs can be used and we encourage our collaborators and students to do so in order to gain experience with the strong and weak points of each program. Each TUTSIM listing is like a set of mathematical equation written in a condensed form; reconverting to usual mathematical writing is immediate (Fig. 2.24/2.25). It can then be converted to BASIC, PASCAL, C, etc., which include other unconventional writings of mathematical equations. In summary, the problems and pitfalls of numerical mathematics are universal.

Today (1990), a book is a data carrier, an alternative to telephone wires and computer diskettes. We have carefully optimised the accessibility, the weak point of books compared to diskettes. So the start is easy, but difficulty increases both with the chapters and with the sections within a chapter.

Fitting modern times, the efficiency of information acquisition by the reader is greatly improved by the personal play with the computer, including the worked examples. Hence we recommend the reader to use his PC as often as possible. It is fun and learning at the same time. In order to make study and play with the examples easier, the reader can order a diskette with the worked examples and a rudimentary form a TUTSIM for a nominal fee.

Concluding, we the authors of Bondgraph books and papers, enjoy the power and beauty of Bondgraphs in teaching, research and consulting since many years. We wish to share our experience and emotions with our readers.

March 1990 Jean U. Thoma

A diskette with the listings of the worked examples and a rudimentary form of TUTSIM can be obtained from:

in Europe: Meerman Automation, Postbus 154, NL 7160 AC Neede
Tel. 0031 5450 93 901, Fax 0031 5450 91 153

in America: TUTSIM Products, 200 California AV #212,
Palo Alto, CA 94306, USA
Tel. (415) 325 4800, Fax (415) 325 4801

Table of Contents

1 Simulation and Graphical System Models 1

1.1 Setting the Stage 1
1.2 Networks for Engineering 4
1.3 Mixed Introductory Topics 6

2 Bondgraphs as Networks for Power and Signal Exchange 9

2.1 Word Bondgraphs and Main Bondgraph Variables 9
2.2 Bondgraph Standard Elements 11
2.3 Computation Marks, Power and Causality 19
2.4 Simulation from Blockdiagrams and Bondgraphs 27
2.5 Nonlinearities and Block-Bondgraphs 30
2.6 Multiport Bondgraph Elements 33
2.7 Junction Structures and Transformations 39

3 Simulation and Design of Mechanical Engineering Systems 41

3.1 Opening of the Application Chapters 41
3.2 Mechanical Systems with Ground Reference 41
3.3 Systematic Writing of Mechanical Bondgraphs with Moving Reference Points 44
3.4 Slip, Clutches and Electric Induction Motors 50
3.5 Transmissions and Efficiencies 54

4 Simulation of Fluid Power Systems and Hydrostatic Drives 57

4.1 Hydrostatic Power in Fluid Flow 57
4.2 Valve Controlled Actuators and Servomotors 64
4.3 Circuits and Hydrostatic Transmissions 68
4.4 Special Components and Devices 71

5 Electrical Circuits, Drives and Components ... 80

5.1 Electric Circuits and Components ... 80
5.2 Electric Motors ... 82
5.3 Capacitors, Inductors, Actuators and Real Transformers ... 89

6 Computational Overview, Practical Procedures and Problems ... 95

6.1 Programs for Simulation from a Bondgraph ... 95
6.2 Linear Control Theory, Eigenvalues and Calculating Step ... 96
6.3 Stability Preview and Steady State Bondgraphs ... 99
6.4 Eigenvalue Estimation and Dynacheck ... 103
6.5 Causality Problems and Algebraic Loops ... 108
6.6 Compact Parameters and Paracheck ... 115
6.7 Causal Loops and Inspectional Analysis ... 119

7 Applications to Thermodynamics, Chemistry and Biology ... 123

7.1 Meet and Love Entropy ... 123
7.2 Convection, Matter Flow and Solar Energy ... 128
7.3 Networks in Chemistry and Physics ... 135

8 Selected Questions ... 139

8.1 Multibonds, Robotics and Circuit Theorems ... 139
8.2 Bondgraphs for Elements Distributed in Space ... 144
8.3 Shocks, Collisions and Hysteresis ... 150
8.4 Analogies, Linear Graphs and Portality ... 155
8.5 The Problem of Alternating Current Bondgraphs ... 162

9 Further Worked Examples ... 165

9.1 Hydraulic Examples ... 165
9.2 Thermofluid Examples ... 174

Appendices ... 177

A1 Organization of Bondgraph Courses ... 177
A2 Note on Bondgraph Standards and Graphics ... 179
A3 Table of the Main Bondgraph Elements ... 181

Glossary of Frequent Terms in English, French, and German ... 183
References ... 185
Subject Index ... 187

1 Simulation and Graphical System Models

1.1 Setting the Stage

In this book we are concerned with the efficient simulation of engineering systems, essentially machines, devices or interconnected apparatus. The purpose is to get a quantitative understanding of their operation, improve their performance by variation of parameters and to find possible critical points before the actual system is built at a high expense. **Our great tool is the modern microcomputer on which engineers wish to work whenever they have a new idea.**

More precisely, we want to calculate the variables of an engineering system and their dynamics from a "mathematical model"

- With a PC (personal computer)
- Dominate the PC and the computation
- See what is calculated

Although we are intimate with the personal computer our main interest lies in real engineering.

1.1.1 Problem Areas

In spite of the power of modern microcomputers, different problems remain.

- Writing and understanding the model
- Inspiring and training our intuition
- Avoiding programming errors
- Escaping the traps of numerical mathematics

In order to overcome these handicaps we develop a new approach which can be called realistic simulation:

- Close contact with engineering
- Graphical system representations
- Efficient with the engineers and scientists time
- Reliable, error free results

We should always remember Thoma's theorem: To err is human, but for complete non-sense one needs a computer.

The keypoint of realistic simulation is the use of graphical system representations which already appeal to western engineers. They should be much more impressive and helpful to Japanese engineers with their familiarity of ideographs from writing.

Among graphical system representations Bondgraphs are most helpful in this context; hence this book is also an introduction to Bondgraphs and their application to engineering simulation on microcomputers.

Systems representations in general range from engineering drawings to sets of equations as shown on Fig. 1.1 in an increasing abstraction from right to left. The central part of the range runs from circuits, familiar in electric and hydraulic engineering, to block-diagrams from control engineering. Bondgraphs are in the middle as a mixture of both with most of the advantages and none of the drawbacks.

We conclude this opening with an example of a practical engineering simulation:

Figure 1.2 shows at the top a Bondgraph and below schematic a large radar antenna. It has an electric AC-induction motor, a slip clutch as torque limiter followed by various inertias, frictions and compliances, all representing a structure with a mass of 135 tons and a height of 68 meters. This antenna was

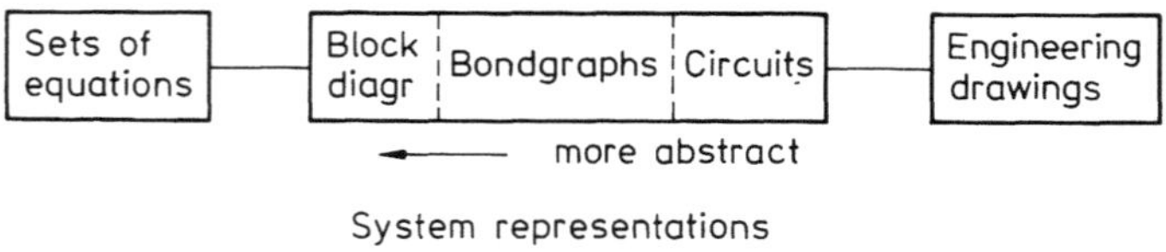

Figure 1.1 System representations ranging from engineering drawings to sets of equations. Abstraction increases from the right to the left. In this book we are mainly concerned with the graphical representations in the middle, particularly with Bondgraphs of course, and their relation to Blockdiagrams and circuits.

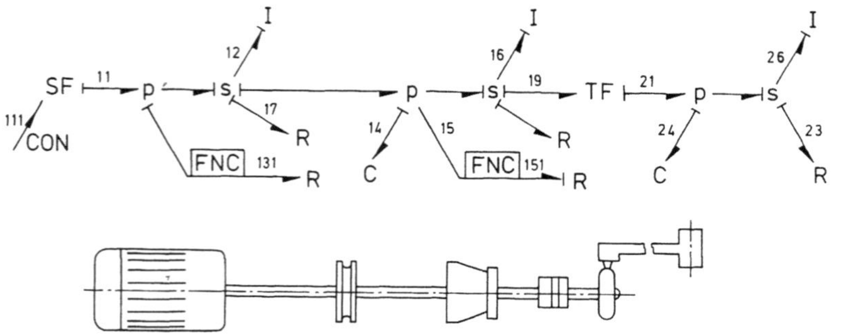

Figure 1.2 Example of the kind of simulations treated in this book: A large radar antenna with electric motor and slip clutch. Schematic below, Bondgraph at the top.

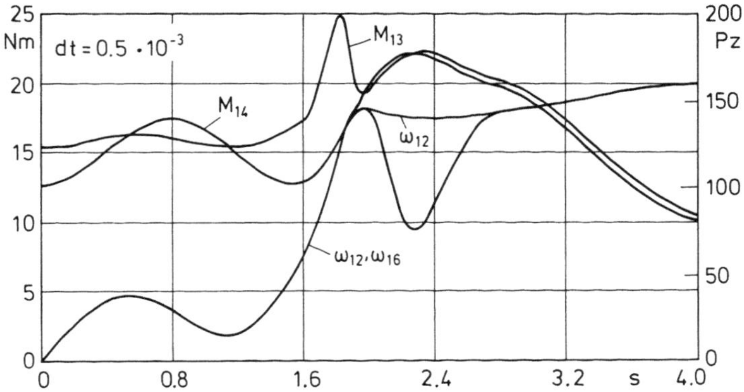


```
  0.0000          1 CON                          ;Rotary Antenna Sottens
  0.0000          2 CON                          ;File Bonsil1.sim
157.0000         11 SF      111                  ;synchro frequency
  0.0480000      12 I        13    -14    -17    ;I of indu motor
  0.0000
 10.0000         13 R       131                  ;slip resistor
  0.0100000      14 C        12    -15    -16    ;shaft compliance
  0.0000
                 15 FNC      14
   1  -30.0000     -225.0000
   2  -15.0000       -1.0000
   3   15.0000        1.0000
   4   30.0000      225.0000
  0.0060000      16 I        14    -18    -19    ;I gear reducer
  0.0000
  0.0031800      17 R        12                  ;bearing friction
  0.0031800      18 R        16                  ;bearing friction
  0.0041800      19 TF       24                  ;gear ratio 1/239
  0.0041800      21 TF       16
690.0000         23 R        26                  ;rail friction
 26.600E-06      24 C        21    -26           ;coupling
  0.0000
  7.800E+03      26 I        24    -23           ; reflec I o antenna
  0.0000
  1.0000        111 CON
                131 FNC      11    -12
   1   -7.5000       -1.0000
   2    0.0000        0.0000
   3    7.5000        1.0000
   4   23.0000        2.4000
   5   32.0000        2.5000
   6   42.0000        2.4000
   7   60.0000        2.0000
   8   91.0000        1.0000
   9  140.0000        0.6000000
  10  157.0000        0.6000000
```

Figure 1.3 Listing and simulation of startup of the radar antenna of Fig. 1.2 with TUTSIM. There is the peak of motor torque M12, typical for an AC-induction motor (compare Fig. 3.16). Then the slip clutch R15 operates temporarily to prevent high torques on the structure, as can be seen by the difference of ω 12 and ω 16. A Bondgraph will help to write and understand the listing and also the simulation results.

overhauled in Switzerland in 1987 and the simulation served to check whether it behaved as expected.

The problem is to write the computer input which in the case of the TUTSIM program leads to the listing of Fig. 1.3 and to the simulation below. The listing could, in principle, be written from the system equations, but it is much more pleasant to go from the schematic to the Bondgraph and then directly to the computer without bothering to write equations at all. This is what we set out to do in our book.

1.2 Networks for Engineering

Having set the stage for realistic simulation from graphical system models, we examine here different network representations, excluding such things as engineering drawings. A network is essentially a set of lumped elements connected by featureless means such as electric wires, pipes or shafts. Featureless means that the connections have no properties of their own like resistance, inductance etc; these can be added as small lumped elements as needed.

Another approach is to consider all relevent dynamic properties distributed in space interacting in very small, infinitesimal space elements. This leads to partial differential equations whilst networks are governed by much simpler ordinary differential equations. Good examples are the Maxwell equations for distributed elements and electric circuits for lumped elements. We know that the circuit approach is very powerful as long as the wave length of disturbances is small compared to the physical size of the systems. Disturbances are electromagnetic, fluid pressure waves or mechanical vibrations (noise) resp. Representing distributed systems by lumped Bondgraph elements is an important extension to be treated in sec. 8.2.

Bondgraphs are a network representation with lumped elements and connections called bonds which also carry computational marks as we shall see. Further, the blockdiagrams of engineering are a similar network, although with slightly different conventions.

Naturally these network representations are related, as shown in Fig. 1.4: A Bondgraph is like a cross-breed of a circuit and a blockdiagram. Main difference is that the circuit uses a picture code for elements and the Bondgraph a letter code (Fig. 1.4 below). The related linear graphs will not be treated in this book. To give an example, Fig. 1.5 contains at the top a simple electric circuit, in the middle the corresponding Bondgraph and below a mixed mechanical/hydraulic system, namely an IC (internal combustion) engine driving a hydraulic pump for fluid power applications. In this sense, a Bondgraph is a network with universal elements representing the common structure of inter-disciplinary systems with the same set of symbols; they are letter symbols for an easy passage to the computer.

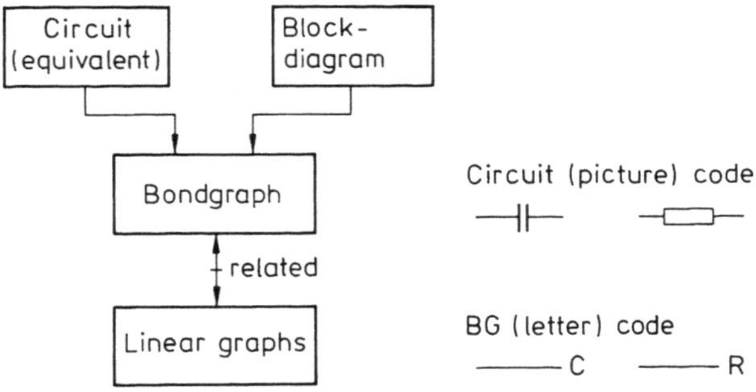

Figure 1.4 Relation between circuits, blockdiagrams and Bondgraphs. A Bondgraph is like a cross-breed of circuits and blockdiagrams and brings out the best properties of both parents; it uses a letter code for the elements whilst circuits use a picture code.

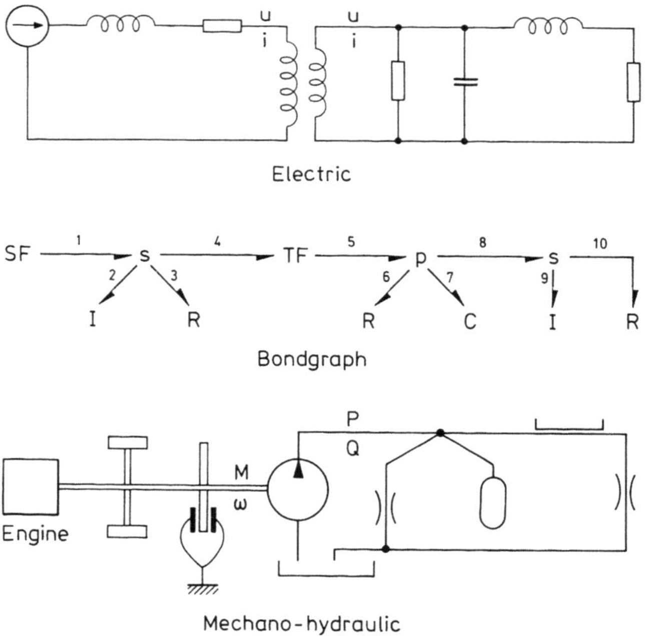

Figure 1.5 Example of a graphical representation with an electric circuit on top, a Bondgraph in the middle and a drive of a hydraulic pump below. The Bondgraph represents both systems with the same symbols.

Our procedure for realistic simulation is as follows:

1. Write a Bondgraph without computational marks
2. Add power directions
3. Add causalities

Points 2 and 3 represent the marks for easy passage to the microcomputers, as will be explained in chapter 2. Applying these marks is called augmenting the Bondgraph and this explicit facility is one of the pillars of their great strength.

1.3 Mixed Introductory Topics

1.3.1 Circuits and Equivalent Circuits

Circuit elements in electrical engineering were originally pictures of real elements or components and the connections were real electric wires; such circuits are today used for the practical building of electric systems from hardware components. For study and quantitative determination of the properties it became convenient to split the real, perhaps complex components into several fictitious parts with simpler characteristics. Thus one obtains an equivalent circuit that has no more (topological) similarity with the real layout.

Figure 1.6 exemplifies the real circuit symbol of an electric battery (left), its equivalent circuit with an ideal voltage generator (called effort source [SE] in Bondgraphs) in series with a resistor representing the internal resistance center. The right part of Fig. 1.6 shows the voltage/current characteristics where the generated voltage is independent of the current drawn. The voltage drop with current is modelled by the internal resistance; in practice its value can be determined from measuring such characteristics.

Two further points become apparent on Fig. 1.6

1. The characteristic can be non-linear outside the range of interest, in particular the voltage might drop much faster at high currents as shown by the broken lines. Within the linear range the series combination of generator and resistor is good.
2. All real components have operating limits which are not part of their performance or control characteristics. On the example of Fig. 1.6, the broken

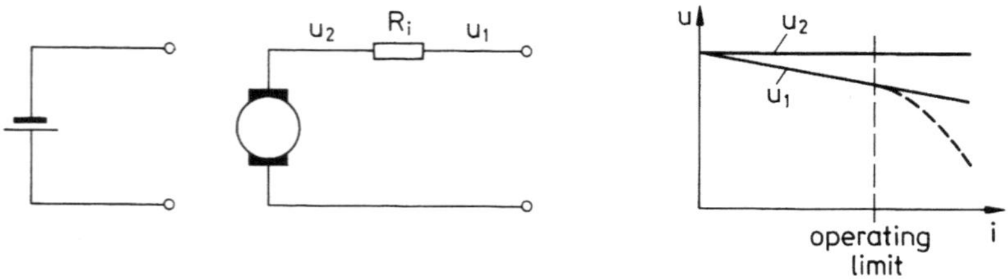

Figure 1.6 Real circuit symbol of an electric battery (top left), equivalent circuit with voltage generator and internal resistance (top right) and voltage/current characteristics below. Operation limits (for current) are determined separately from required life and reliability and indicated by the broken vertical line. Real characteristics may be non-linear outside the range of interest.

vertical line is a current limit. If exceeded, the battery will be damaged or even destroyed. The operating limit is somewhat arbitary, it depends on required life, reliability and often short overloads are admitted.

1.3.2 Blockdiagrams and their Inversion

Blockdiagrams of control engineering are made of the following elements:

1. Blocks with one output which is a function of one or more inputs.
2. Connections where a signal travels in a definite direction as indicated by the (full) arrow.
3. Mixing points where the output is the sum or difference of two or more input signals.
4. Branch points where the signals split up without changing in value.

Figure 1.7 shows at the top a simple example with 2 blocks having gains F and G, connections carrying signals s1 to s4, a mixing and a branching point. Note that the gains can be linear or not and can contain dynamical elements, delays or be a function of the signal frequency (or of the Laplace operators). The figure also shows the equations, the loop gain and the transfer function between input s1 and output s3 as obtained either by mathematical manipulation or by Mason's rule.

The signals travel in the direction of the arrows, but is such a direction fixed, given by a physical law? No, in Fig. 1.6 below, the signals travel (partly) backwards, that is in the inverse direction. Now the block gains become reciprocal. The loop gain is also reciprocal, 1/FG, but the overall transfer function remains the same (again by manipulation or Masons's rule). Hence, both blockdiagrams can represent the same system. If the gains F or G contain

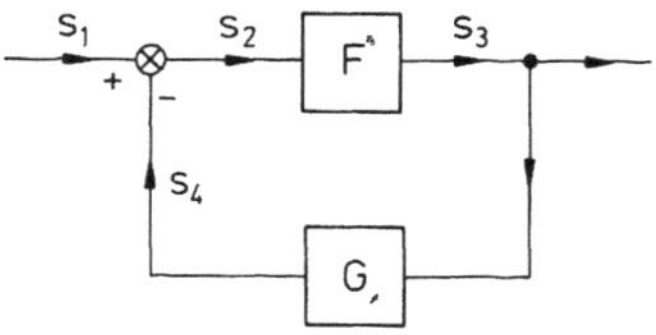

Equations | Loop gain = FG

$s_2 = s_1 - s_4$

$s_3 = F s_2$

$s_4 = G s_3$

transfer function

$$\frac{s_3}{s_1} = \frac{F}{1 + FG} \quad \text{Mason}$$

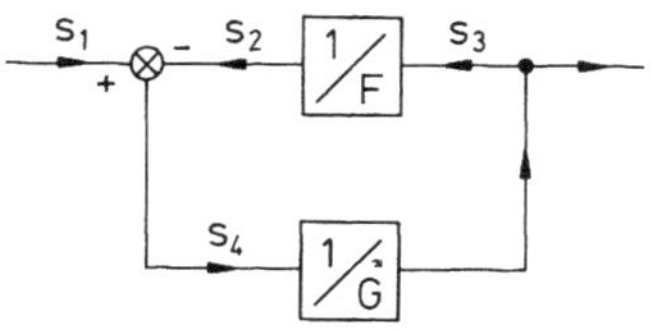

Equations | Loop gain = 1/FG

$s_4 = s_1 - s_2$

$s_3 = s_4 G^{-1}$

$s_2 = s_3 F^{-1}$

transfer function

$$\frac{s_3}{s_1} = \frac{G^{-1}}{1 + F^{-1} \cdot G^{-1}} = \frac{F}{FG + 1}$$

Figure 1.7 Simple blockdiagram with equations, loop gain and transfer function (top) and its inversion (bottom). In the lower part, the gains including the loop gain have become reciprocal but the transfer function remains unchanged, hence the inverted blockdiagram can represent the same system.

derivations or integrations, they will be converted to integrations and derivations, respectivly, as represented correctly by manipulation of the Laplace operator.

The curious change of Fig. 1.7 is an inversion of a blockdiagram. Signals will travel backwards and causes and effects are permuted (mutually interchanged). This is called change of causality and plays a fundamental role in Bondgraphs and in computer simulation, both on microcomputers and main frames in general. As stated before, the systematic and graphical causality assignment is a pillar of strength of Bondgraphs.

2 Bondgraphs as Networks for Power and Signal Exchange

2.1 Word Bondgraphs and Main Bondgraph Variables

For simulations of any complexity it is good to use a systematic way of building the model in small steps. A first step is to write a word Bondgraph; it contains words, not standard symbols for the main items (a subgroup) and normal bonds for power and signal exchange.

Figure 2.1 contains an example of a motor car with engine, transmission, wheels and road. It already displays a lot of information, namely nature of the variables M/ω (torque/rotation frequency) and $F/\dot{x}$ (force/speed) between the wheels and the road. Further the half arrows indicate that the normal flow power goes from the engine to the road. Finally both engine and transmission are influenced by the control signal in the line with the (full) arrow, familiar from blockdiagrams of control engineering (sec. 13).

For simulation, the next step is to replace the words by standard elements which contain precise mathematical/functional relations. How many of these are needed depends on the desired accuracy or maximum signal frequency and is an important modelling decision. We also want to know the exact meaning of the half and full arrows.

As apparent from Fig. 2.1, each bond contains 2 variables, the product of which is a power. This is always so in a true Bondgraph and table 2.1 shows in its upper two lines effort and flow variables for all disciplines. The words effort and flow designate voltage and current like variables in general. The table also contains the usual letter (algebraic) symbols which we shall normally use in this book.

Furthermore, table 2.1 contains in its lower part variables derived from effort and flow, namely momentum as time integral of effort and displacement as time integral of flow. Both statements are definitions and displacement and momentum are fundamental in many parts of physics. The last two columns of table 2.1 refer to thermal and chemical networks and seem to have more importance in physics than in practical simulation (so far).

In most practical simulations heat flow is taken as flow variable and temperature as effort. In such a Bondgraph, the product of effort and flow is no longer a power and has no meaning; thus this is a pseudo-Bondgraph. Table 2.2 shows these pseudo-Bondgraphs variables and another pair, concentration and

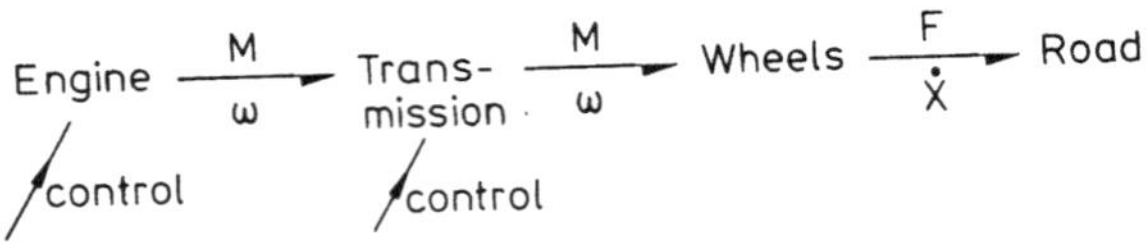

Figure 2.1 Word Bondgraph for a motor car drive. Each word represents a main item or subgroup, and the lines between are the bonds. Letters indicate the kind of variables, torque/rotation frequency or force/speed and the half arrows indicate the normal direction of power flow. The lines with the full arrows are connections in the sense of control engineering.

Table 2.1 Table with the effort and flow of true Bondgraphs in different disciplines designated by the usual letter symbols. The lower lines give examples of momentum and displacement defined as the time integrals of effort and flow resp.

General	Electric	Hydraulic incompressible	Mechanical rectilinear	Mechanical rotary	Thermal	Chemical
effort	voltage	pressure	Force	Torque	Temperature	Chem Pot
e	U	p	F	M	T	μ
flow	current	volume flow	speed	rota frequ	Entropy flow	molar flow
f	i	$Q, \dot{V}$	$\dot{x}$	ω	$\dot{S}$	$\dot{n}$
Momentum	electric momentum	Hydraulic momentum	momentum	angular momentum	?	?
p	pel	p	p	L		
Displacement	elect	volume	position	angle	Entropy	molar mass
q	q	V	X	ϕ	S	n

Table 2.2 Table with variables of pseudobondgraphs, where the product of effort and flow is not a power and has no physical significance. For practical simulation the temperature/heat flow pseudobond is more convenient as long as there is no thermo-mechanical power conversion.

	Thermal	Chemical
effort	Temperature	Concentration
	T	C
Flow	Heat Flow	molar Flow
	$\dot{Q}$	$\dot{n}$
Momentum	?	?
Displacement	Heat	molar Mass
	Q	n

molar flow in chemical kinetics. We shall return to thermal and chemical Bondgraphs, true and pseudo in chapter 7. Here it should only be noted that the pseudo variable heat flow is more convenient for simulation as long as no thermal/mechanical energy conversion takes place.

Returning to table 2.1, force is taken as effort and speed as flow, also rotary torque and rotation frequency. This is called the force/effort analogy used in this book. There also exists the force/flow analogy, with force as flow and speed as effort. We shall return to them in chapter 8. Meanwhile most Bondgraphs are written in the force/effort analogy.

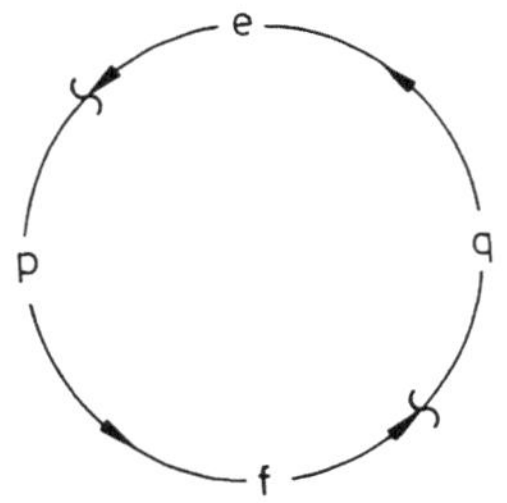

Figure 2.2 Circular diagram or carrousel for the relation between flow, displacement, effort and momentum, with integrations is the CCW sense. Other relations will take in the oneports (elements with only one bond) explained in section 2.2.

Our four kinds of variables, flow, displacement, effort and momentum can be arranged on a circle in Fig. 2.2. It displays the integrations required to go from flow to displacement and from effort to momentum in a CCW (counter clockwise) rotation. Left open are the relations between displacement and effort and between momentum and flow. They take place in one-port elements; such elements have only one bond and will appear at the end of sec. 2.2

Let it already now be noted that the momentum is not, definitely not, defined as mass times speed. This is a relation between momentum and flow and belongs to the lower left of Fig. 2.2 where it is called a constitutive equation. With very high speeds the constitutive equation changes as we know from Einstein's relativity, but has no importance to practical simulation at the present time (1990).

2.2 Bondgraph Standard Elements

The next step after the word Bondgraph of sec. 2.1 is to replace the words by standard Bondgraph elements, each one representing a definite effect or action in our system. An element is more than a mathematical relationship and has three levels of description which are a central guide line of system simulation by Bondgraphs:

1. Constitutive equation
2. Energy and power relations
3. Reversibility or not (from thermodynamics)

For each element we shall study the behaviour on the three levels; levels 2 and 3 are more crude but help to prevent simulation errors in practice.

A possible classification of elements is by the number of bonds as one-ports, two-ports, three-ports and multiports. We shall use this classification here to bring order into our element description.

2.2.1 One-Ports

Elements with only one bond are generalized resistors, inductors and capacitors, R-elements, I-elements and C-elements resp. Generalization is from electricity

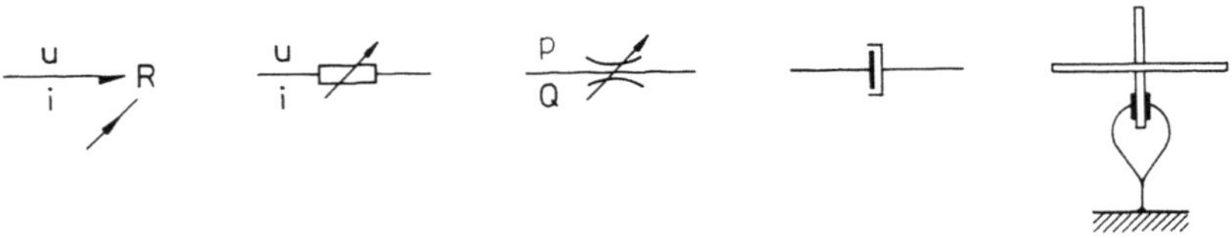

Figure 2.3 Resistance or R-element, top left as Bondgraph symbol, followed by electric, hydraulic, rectilinear and rotary mechanical system symbols; the last should resemble a disk brake. Some times the resistance value can be controlled as indicated by the oblique connection to the Bondgraph symbol and by the cross arrow in the electric and hydraulic case.

into other domains and such elements appear frequently. Additional elements are generalized voltage and current sources SE and SF which normally supply power.

2.2.2 R-Elements

Figure 2.3 displays as the left the resistance or R-element as a Bondgraph symbol, followed by internationally standardized electric and hydraulic symbols. The second line contains a rectilinear resistor symbol used in mechanical engineering text books resembling a dashpot and followed by a Thoma symbol[1] for a rotary resistor resembling a disc brake of automobiles. Dashpots and brakes are in fact real components installed for damping or braking action.

All elements can be controlled by or depend on other variables like the position of a control rod. The Bondgraph element is thus attacked by a signal connection with full arrow (oblique on Fig. 2.3). On the other hand, circuit symbols receive a cross arrow to indicate controlability according to the norms. The circuit symbols are simplified pictures of the real elements that can not be transferred easily to the computer, but the letter symbols like our R can be entered through the keyboard in most programs. The same is true for the following C-elements and I-elements.

Referring to our three levels of description of R-elements:

1. The constitutive equation is between voltage and current, effort and flow in general and may be linear or not.
2. Power always enters and never returns: The R-element dissipates and produces heat flow, or more precisely entropy-flow (Fig. 7.4). Neglecting heat production it is a power sink, including it becomes a power conserving element.
3. R-elements are irreversible.

[1]Such symbols are not standardized but were found convenient by the author

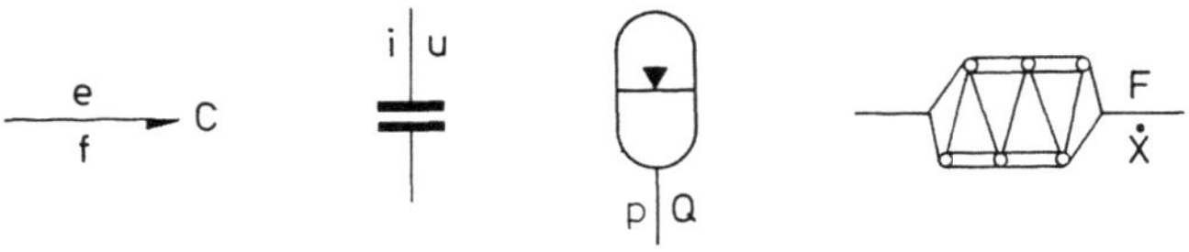

Figure 2.4 Generalized capacitor or C-element with electric capacitor, hydraulic accumulator and spring as circuit examples. These elements are energy conserving and reversible.

2.2.3 C-Elements

Figure 2.4 contains a generalized capacitor or C-element, at the left as Bondgraph symbol–C. There follow the electric and hydraulic picture symbols, internationally standardized for capacitor and accumulator. At the right there is a picture of a spring. In fact, C-elements model electric capacitors, hydraulic accumulators and linear or rotary springs. These elements may be separate, visible components or implicit in others, like electric line capacitance or torsional compliance of a shaft. Control is similar as before.

The three levels of description:

1. The constitutive equation is between charge and voltage (electric) and between spring displacement and force (mechanical), in general between effort and displacement and independent of time.
2. C-elements are energy, not power conserving. Here power is absorbed when the spring is squeezed, but it is stored inside as energy and changes the state of the element. Returning to the original state (length in the case of the spring) the entire energy is returned through the bond: Energy conservation.
3. C-elements are reversible.

2.2.4 I-Elements

Generalized inductors in Fig. 2.5 comprise the electric inductor, fluid inertia in hydraulics, for which no symbol is universally accepted, but a Thoma symbol appears on Fig. 4.8. Rectilinear mechanical inertia is included and pictured as a sliding block, rotary inertia as a fly wheel; this is essentially a big wheel used to

Figure 2.5 Generalized inertia or I-element with electric inductor, mass action and flywheel as examples. The latter is an artificial I-element to smooth rotation.

smooth rotation of steam and IC-engines, an artificial I-element. Hydraulic I-elements are the mass action in fluid pipes, a very important effect for fast systems but as stated no suitable circuit symbol exists. The three levels of description are similar to the C-elements:

1. Constitutive equation is between momentum and speed, namely speed equals momentum divided by mass. This is equivalent to Newton's law where acceleration force is mass times acceleration (or acceleration torque equals moment of inertia times rotary acceleration).
2. I-elements are energy storing like C-elements.
3. I-elements are reversible.

Both C-elements and I-elements are also called energy storing elements, or energy storers. This term includes the multiport C-fields and I-fields below in sec. 2.6.

2.2.5 Carrousel of Variables and Elements

As stated before the essence of R-, C-, and I-elements is not that they are linear, and many are not, but to which variables they relate. In order to more easily remember we can develop from the circular diagram 2.2 the more complete diagram on Fig. 2.6, called Thoma's carrousel of variables and elements[1]. We can see there that the C-element makes a functional relation between displacement and effort, the I-element between momentum and flow, and the R-element between effort and flow. This latter relation has no preferred direction as indicated by the opposed arrows, we can say indifferently voltage is current times resistance or current is voltage divided by resistance (referred to as resistance and conductance causality, resp. see sec. 2.3).

However, there is a preferred direction for the energy storing elements (C-elements and I-elements): rotating CCW on Fig. 2.6, we encounter only integrations. This is important since computers like integrations and hate derivations. So the art of simulation is to program such that only integrations are

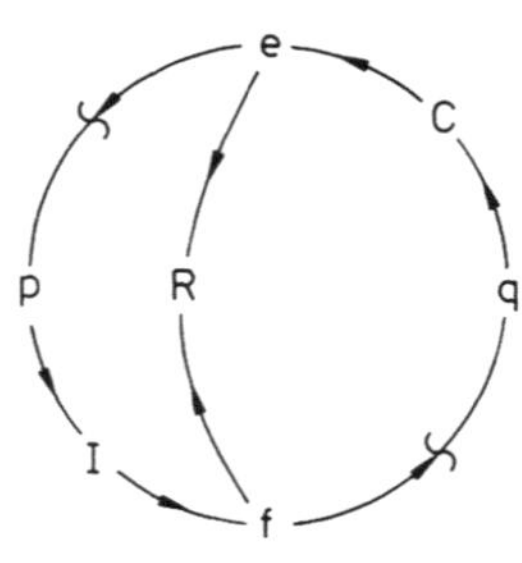

Figure 2.6 Thoma's carrousel displaying the action of I-, C-, and R-elements that relate momentum and flow, displacement and effort and flow respectively. Relating these variables makes the essence of the elements, whether they are linear or not.

[1]It is based on an older representation of Henry Paynter, called Tetrahedron

required, for which Bondgraphs offer a systematic help by the causality assignment procedure (sec. 2.3).

2.2.6 Effort and Flow Sources

The voltage and current sources, well known in electrical engineering, are usefully generalized into effort and flow sources on Fig. 2.7; they have the Bondgraph letter symbols SE and SF at the left and electric symbols at the right.

Effort sources do:

1. supply and effort independently of flow required of flow required in response to a control signal
2. normally supply power to the system but can absorb power when driven in reverse
3. are reversible in the sense of thermodynamics.

Flow sources are similar with effort/flow interchanged. Sources can be controlled as indicated by the oblique control lines. The meaning of the cross-stroke (selected causality) will be explained in sec. 2.3.

Figure 2.8 shows that effort is independent of flow for the effort source (horizontal line) and flow independent of effort for the flow source (vertical line). The position or value of the vertical (horizontal) line for the constant effort (flow) can be controlled in general. These lines can extend beyond the zero lines (coordinate axis). In this case the source is reverse driven and returns power to the system. Most real sources admit such a condition of operation.

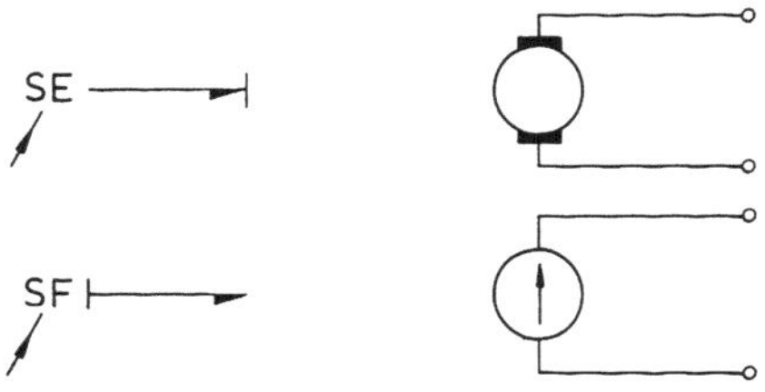

Figure 2.7 Effort and flow sources with electric circuit examples. Effort source gives effort or voltage independent of flow and flow source gives current or flow independent of effort. Such sources are frequent also in hydraulics and mechanics; they can be controlled as indicated by the oblique connections.

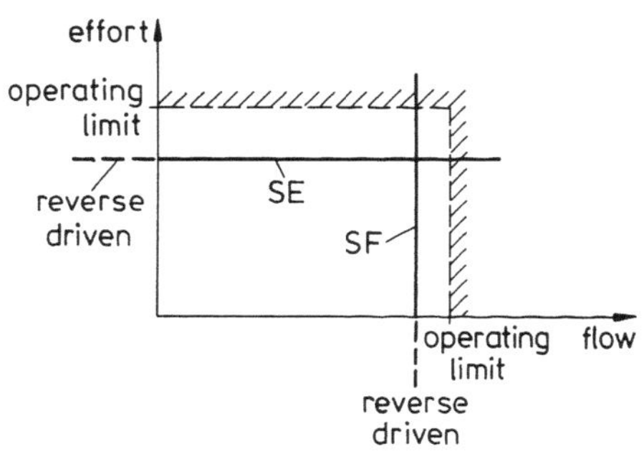

Figure 2.8 Characteristics effort/flow of sources: Horizontal for effort source, vertical for flow source. Shown are also the operating limits which must not be exceeded to avoid damage of real components. Normally the control system prevents this.

Figure 2.8 also contains the operation limits of the source in broken lines, the maximum admissible flow or effort; these limits depend on required life and reliability. They are not inherent in the source and its characteristics; rather, the control system must avoid passing the limits to avoid damage. In particular if a pure effort source feeds into very low load excessive current is drawn. This is called a short circuit and special circuit breakers (fuses) are often installed to disconnect the load, in this case to prevent destruction.

In a real (electric) circuit, voltage will decrease somewhat with the required current, linearly in an element like a battery. This can be represented by an effort source with a resistor in series or by a flow source with a resistor in parallel; in the linear case both resistances are equal and are called the internal resistance. These representations are called Thevenin and Norton equivalent circuit, respectively and their equivalence is the theorem of Thevenin.

Our Fig. 2.7 also shows electric circuit symbols for the sources, such sources also appear in other disciplines, as will be illustrated by the following examples:

Hydraulics: a constant speed driven pump is a flow source, also a circuit with a flow control valve. A relief valve or pressure regulator added in parallel changes it into an effort source (see sec. 4.4).

Mechanical: Gravity force is an effort source, and a large and soft spring approximates it (although it is really a C-element). A flow source is an old fashioned overhead transmission shaft; further an electric DC motor with constant excitation and supplied from a voltage source (sec. 5.2). In both cases shaft frequency is independent of required torque.

The above 5 elements, including the two sources, are all the one-ports we require for the simulation of even complex systems. They will be connected by the other elements, namely two-ports, three-ports and partly multiports, as we shall see in what follows.

2.2.7 Two-Ports

There only exists two kinds of two-ports (excluding the multiport fields of sec. 2.6 with two-ports) namely the transformer and the gyrator.

The transformer is best known from electrical engineering and has the symbol TF for fixed ratio and MTF for variable ratio that can be modulated (varied) by a control signal: it then becomes a Modulated Transformer.

Figure 2.9 shows some examples, namely an electrical transformer, a mechanical gear reducer, a hydrostatic pump and a lever for small rotations.

The three levels of description for a transformer are:

1. Constitutive equation: Both efforts are proportional and also both flows (see sec. 2.3 for formulae).
2. Transformers conserve power.
3. Transformers are reversible.

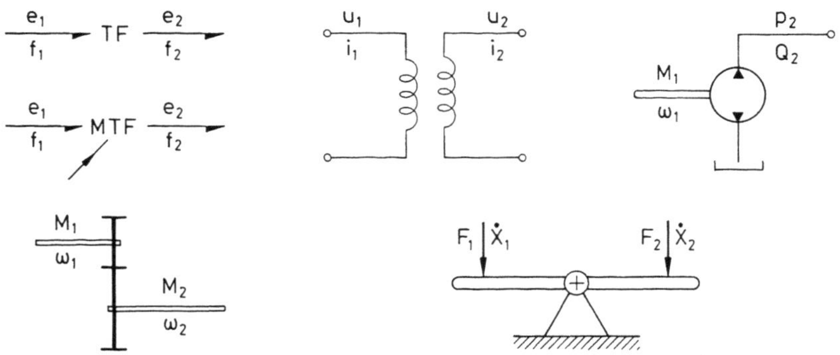

Figure 2.9 Transformer (top left) and modulated transformer (bottom left) as Bondgraph symbol with corresponding electric, hydraulic (pump), rotary mechanical (gear reducer) and rectilinear mechanical (lever) examples. Circuit symbols can receive a cross arrow to indicate variable ratios similar as in Fig. 2.3.

Power conservation means that at each instant of time the entering power equals the leaving power. Naturally this requires two (or more) bonds and a one-port cannot be power conserving. As a consequence the constant of proportionality in the constitutive equations is equal for flow and for efforts if these are expressed in compatible measuring units such as SI units.

An intriguing question: Transformers in the sense of Bondgraphs also work for DC (direct current), that is with zero frequency, f.i. in a gear reducer for constant rotation. A real electric transformer only works with an AC (alternating current) and gives a short circuit on DC. Why is this so? The answer to the riddle will appear in sec. 5.3.3.

From the physical point of view, there are two kinds of transformers:

1. The class transformer that connects different classes or physical domains of variables, like mechanical rotary and hydraulic variables in a hydrostatic machine (Fig. 4.15).
2. The impedance transformer which connects variables of the same class and changes only the effort/flow ratio or impedance: a gear reducer is an example.

Gyrators are similar two-ports and also called over crossed transformers, since effort in one bond is proportional to the flow in the other. The symbol is GY or MGY for Modulated GYrators as shown with the important electric example on Fig. 2.10, the DC electric motor. Gyrators are power conserving and reversible and the same remarks apply.

In small motors the normal engineering description is that torque is proportional to armature current; consequently the reflected voltage (back EMF) is proportional to the rotation frequency. Both conversion parameters are equal by power conservation; they are constant in the case of permanent magnets for field excitation. This magnetic field can also be excited by a field coil, whence the

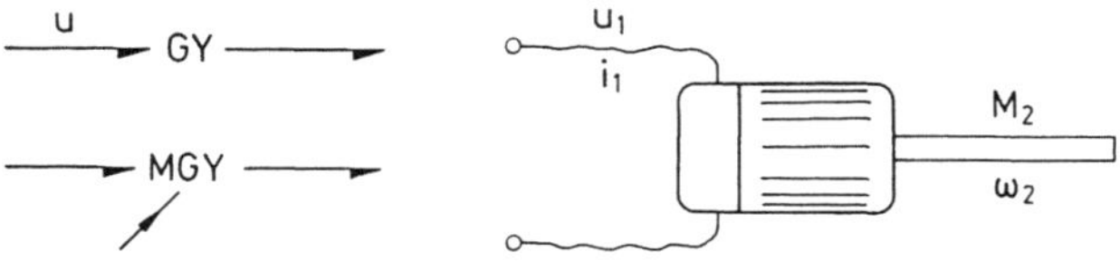

Figure 2.10 Gyrator and modulated gyrator with electric motor example.

conversion parameter becomes proportional to the field current and the motor behaves as a MGY.

2.2.8 Three-Ports: Parallel and Series Junctions

Junctions normally connect three bonds, but exceptionally also two or four or more bonds; they are called three-ports for convenience. We have only two kinds, the parallel and the series junction illustrated on Fig. 2.11. They conserve power and are reversible.

Parallel junctions have equality of efforts whilst the flows sum up to zero if power orientations are taken positive toward the junction. The junction itself is designated by the letter p (or the numeral 0).

Series junctions have equality of flows and the efforts sum up of zero with the same power orientation; they are designated by the letter s (or the numeral 1).

In electric and hydraulic systems the junctions correspond to parallel and series connections of wires or hydraulic pipes. We thus prefer the p/s notation for its mnemonic qualities, although originally the 0/1 notation was introduced.

Junctions can also have 4 or more bonds, whence equations in Fig. 2.11 need to be adjusted. Junctions with only 2 bonds are an identity with suitable power orientation and are occasionally used to facilitate labelling in TUTSIM.

At this point we have described all fundamental Bondgraph elements and we conclude the section with a simple example:

Figure 2.12 shows an electric circuit and a Bondgraph with a current source, capacitor C2 in parallel to a resistor R3 and an inductor I4 in series (in

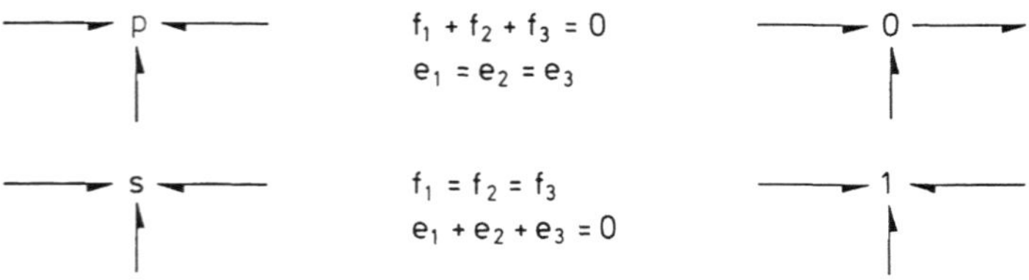

$$f_1 + f_2 + f_3 = 0$$
$$e_1 = e_2 = e_3$$

$$f_1 = f_2 = f_3$$
$$e_1 + e_2 + e_3 = 0$$

Figure 2.11 Parallel and series junctions with equations: In parallel junctions efforts are equal and flows sum up to zero, in series junctions flows are equal and efforts sum up to zero. An alternative notation is 0/1 (zero/one junction).

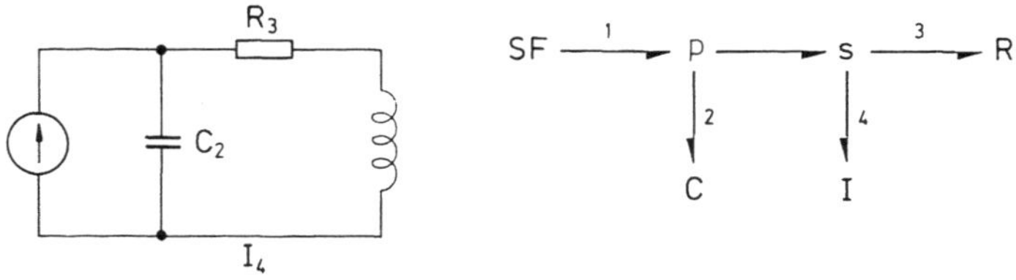

Figure 2.12 Simple electric circuit and corresponding Bondgraph. Bonds are labelled and elements are defined by their bond number as C2, R3 etc.

Bondgraphing we use the letter I instead of the usual L for electric inductors). We note also from Fig. 2.12 at the right that the bonds are labelled, consecutively or not, and the elements defined by their bond numbers. Having selected a labelling scheme from the Bondgraph, the elements can be retro-labelled on the circuit. Later (sec. 5.1) we show how to write Bondgraphs systematically from electric circuits. Now it is sufficient to note how the parallel and series arrangements in the circuit become p-junctions and s-junctions in the Bondgraph.

A slightly more complex example appeared in Fig. 1.5; here the TF would be labelled TF 45. Such writing is the principle of transforming Bondgraphing structure into computation by the CAMP and ENPORT programs. TUTSIM uses a different approach (sec. 2.4).

2.3 Computation Marks, Power and Causality

In this section we treat the equations of the Bondgraph elements presented in sec. 2.2. Here the half arrow for power will play an important role and we shall introduce causality, a central notion in computer applications and programing.

Briefly, the half arrows for power indicate that the product of effort and flow is positive if power travels in this direction. The causal cross-stroke indicates on which side of the bond the effort acts. Both conventions will become clearer by the following examples.

2.3.1 Energy Storing C- and I-Elements

Figure 2.13 refers to a C-element and contains on top in the middle one form of the constitutive equation, that flow (current) is proportional to the time derivative of voltage. At the right it is represented as an element of a blockdiagram, whence s is the derivation operator (Laplace variable). The gain of the block is seen to be sC. On the Bondgraph element at the left we can imagine (and that is a universal picture) that effort travels on the top side towards C and the calculated flow travels backwards. In other words, if the effort acts upon the C-element it responds by a flow, a process called handshaking. This picture is coded (indicated) in the second line as a cross-stroke on the bond near the letter C.

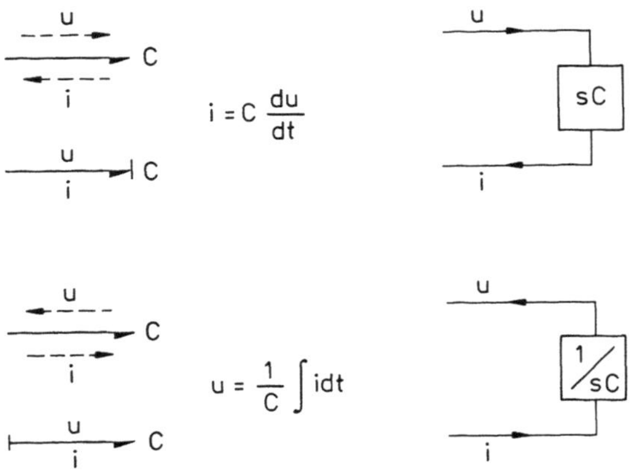

Figure 2.13 Causalities, equations and blockdiagram for C-element on top where effort travels towards the element which responds with a flow. This is derivative causality since flow comes from derivation of effort, as indicated by the derivative operator s in the block. The signals (effort and flow) travel along the sides of the bond as indicated concisely by the causal stroke (second line). The lower part indicates integral causality with flow travelling towards the C-element that responds with an effort. The block now contains the integration operator 1/s.

The third line of Fig. 2.13 contains another form of the constitutive equation where effort is proportional to the time integral of flow. Again at the right there appears a block where the operator 1/s indicates integration, conforming with the equation in the center. At the left there is the Bondgraph element with the flow now travelling towards the C and effort backwards. This is encoded in the bottom line with the cross-stroke on the side away from the C. Note also that the gain of the block is now 1/sC, the reciprocal of the gain in the block of the first line. It looks like an inversion of an element of a block diagram of sec. 1.3 and this is in fact what it is.

Applying cross-strokes is thus a short cut to indicate the direction of action (cause to effect) of the effort and flow in a bond. This direction is quite separate and independent of the direction of flow or current in a real electric circuit, or of fluid in a hydraulic circuit. The process described above is called selection of causality.

Figure 2.13 thus contains 2 possible selections of causality, where the upper part indicates that the output (flow) comes from the time derivation of the input signal (effort). The lower part indicates that the output (effort) results from integration of the input (flow). In other words we have derivative and integral causalities. In computation the crucial (decisive) point is that computers can integrate easily and derivate only with great difficulty: Computers love to integrate and hate to derivate. The essence of good programing is to apply causalities such that only integrations occur. For this we use systematical a causality assignment as detailed below (sec. 2.3.9).

Figure 2.14 shows similar relations for an I-element, where the roles of effort and flow are reversed. We obtain integral causalities when effort attacks the I-element which responds by calculating a flow; derivative causality results with a flow acting on the I-element which responds with an effort. Again we should assign integral causality for easy computation.

The standard form of Newton's acceleration law states that effort equals mass times acceleration (time derivative of speed). It is thus derivative and should be

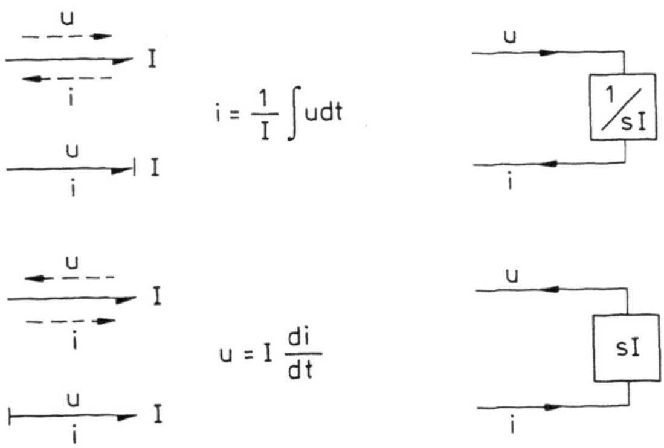

Figure 2.14 Causalities, equations and blockdiagram for an I-element. Here effort is travelling towards the element and flow responds leading to integral causalities; the contrary, flow travelling to the element and effort returning, gives derivative causality. We write here (as in Fig. 2.13) u and i for a concrete example, but generally it is e and f.

replaced by the statement that speed is momentum or time integral of force divided by mass. The curious fact is that humans prefer to think in derivative causality and computers in integral causality. To conclude causality of C- and I-elements, it is always desirable to have them in integral, but not always possible. We speak here of desirable causalities.

2.3.2 Mandatory and Desirable Causalities

At this point we would like to distinguish between the causality rules:

- Desirable causalities are those that lead to integrations on energy storers, essentially C- and I-elements. In certain cases they need to be violated, made undesirable, as we shall see in sec. 6.4
- Mandatory causalities are all other ones as described here, in particular for junctions, transformers and sources. They must be used, otherwise the Bondgraph is incomputable, that is the corresponding equations contradict each other.

2.3.3 Power Directions

The half arrows indicate the direction of positive power flux. Since power is the product of effort and flow, either both are positive or both are negative if power flux is in the direction of the half arrow. If power travels backwards one of them must be negative (but not both). In particular in a C-element if flow enters the effort increases, reflecting the storing of energy. Hence the signs of Figs. 2.13 and 2.14. Here the parameter of the C-element, the capacity, will generally be positive. If not we have a spring with negative slope that can and does lead to instabilities.

Figure 2.15 shows another assignment of power directions, power being positive if leaving the C- or I-elements. The corresponding equations now have

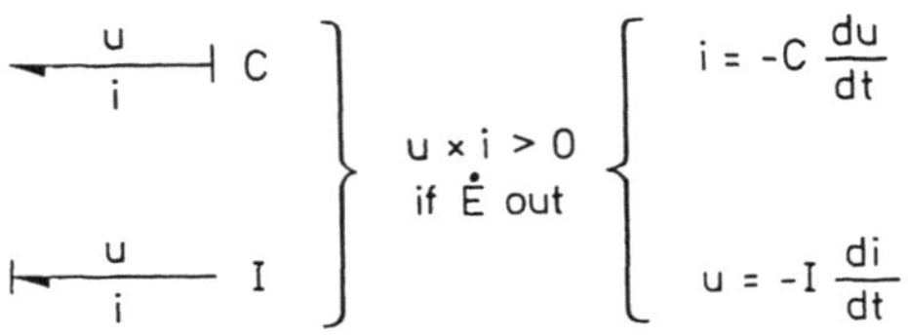

Figure 2.15 Unusual power orientation in C- and I-elements. Here the equation needs the minus signs (or a negative value of the parameter) to represent physical reality in most elements. Contrary behaviour is possible, f.i. a spring, where force decreases with deformation, but easily becomes instable.

negative signs. In Bondgraphing such an orientation is possible but not recommended because of the danger of sign confusion. In standard electric engineering such power orientation is usual. Here the minus sign as shown must be added or the value of the parameter must be negative for stability.

2.3.4 Causalities and equations for R-elements

Figure 2.16 represents the two possible causalities of R-elements. On top flow attacks the R-element which calculates an effort according to the (linearized) equation in the center. At the right it is represented as an element of block diagrams with resistance as gain of the block. This is called resistance causality.

The lower line of Fig. 2.16 represents conductance causality with the cross-stroke near to the R, indicating the effort acts along the bond on the R-element, which replies by a flow. This corresponds to the equation in the center and to the block at the right with the conductance, the reciprocal resistance as gain.

For computation resistance or conductance causality makes no difference in principle, as long as the elements are linear. However, the selected causality indicates in most programs whether resistance or conductance has to be entered as a parameter.

2.3.5 Power and Energy Limits of One-Ports

Power as energy flux and also energy have universal significance, values and signs, whilst the signs of flows and efforts are a little arbitrary.

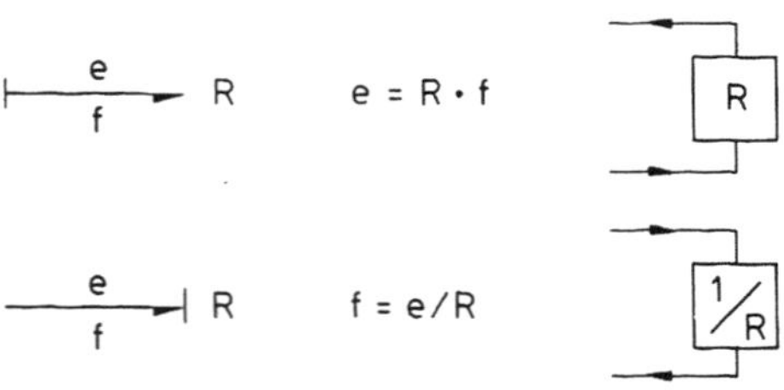

Figure 2.16 Two possible causal selections in an R-element, resistance and conductance causality. Mathematically they are equivalent, but in practice they show whether resistance or conductance value needs to be programed as parameter.

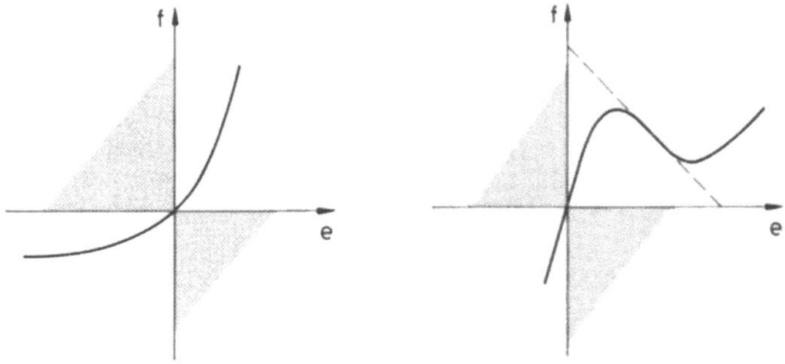

Figure 2.17 Effort/flow characteristic of an R-element in normal power orientation. Since power cannot come out, the shaded quadrants 2 and 4 are forbidden (left). Even a negative resistance, such as of a tunnel diode is only possible in the allowed quadrants 1 (or 3) and over a limited range.

Figure 2.17 refers to a (non-linear) R-element and shows the characteristics, flow as function of effort (conductance causality) in normal power orientation. Power can only enter the R-element and never leave it, hence the second and fourth quardrant are forbidden (effort negative, flow positive or effort positive, flow negative). This prohibition is also called the second law of thermodynamics and imposes that the value of the gain (resistance) be positive.

The right of Fig. 2.17 shows how negative resistances are possible over a limited range as long as the forbidden quadrants are avoided. It is the characteristics of an Esaki (tunnel) diode which often produces oscillations. More precisely the linear part can be represented by an effort source in series with a negative resistor or by a flow source in parallel with the same negative resistor (Thevenin or Norton equivalent circuits).

What happens when an experimental characteristics does not avoid the forbidden quadrants? Well then the component is not a pure R-element but has something inside that supplies power. An example would be the initial current of an old fashioned radio tube, which is really a thermoelectric power converter from the hot cathode to the cold anode.

C- and I-elements have not a power but an energy limit. Figure 2.18 shows a linear C-element with effort proportional to displacement and energy proportional to the square of displacement. Here we have a lower energy limit at zero displacement and effort. Imposing a negative displacement, effort becomes negative and stored energy increases again. From the zero energy level it is impossible to take more energy out, just as in an empty money box no amount of scratching the bottom will produce more money. I-elements are the same, only with flow and momentum instead of effort and displacement.[1]

[1]Note for physicists: Quantum mechanics gives much importance to the zero point energy of harmonic oscillators, which are a I- and C-combination. Perhaps a reader proficient in Bondgraphing and quantum mechanics can be inspired to investigate

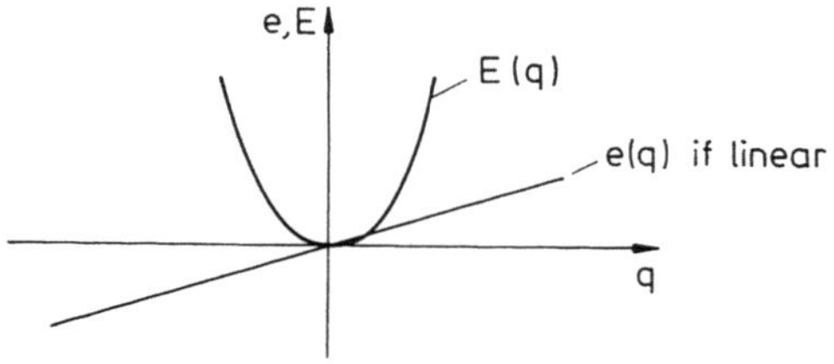

Figure 2.18 Characteristic effort/displacement and stored energy/displacement of a C-element. The stored energy has an absolute minimum (set to zero) for zero displacement and effort. Imposing negative displacement requires negative effort and again positive energy for storing.

2.3.6 Power and Causality for Sources

Normal power orientation is positive out of the source, corresponding to the idea that it supplies power to the system, see Fig. 2.19. Thus an effort source supplies as much flow as necessary to keep effort constant. Possible it has to withdraw flow (and power) if the system wants to overrun and the flow becomes negative. Since effort remains positive, power becomes negative and we have a reverse driven source, sometimes called a sink of power.

Conversely, flow sources impose as much effort to the system as needed to maintain the flow constant. Effort can also become negative if needed to maintain flow constant.

For causality, an effort source always imposes effort, a flow source a flow as shown by the causal strokes on Fig. 2.19. These are mandatory in distinction with the desirable integral causalities of C- and I-elements. Mandatory causality cannot be violated. If a Bondgraph does so, its underlying equations are in contradiction and nothing can be computed: A Bondgraph with violated mandatory causalities is incomputable.

Figure 2.19 Preferred power orientations and mandatory causalities of effort and flow sources.

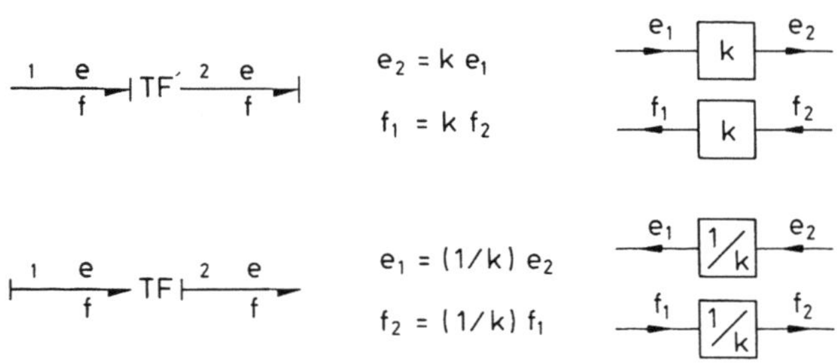

Figure 2.20 Two possible causalities for transformers with causal stroke at left, equation in the middle and blockdiagram at right. Going from one causality to the other, indifferent for computation, requires inserting the reciprocal parameter 1/K instead of K for the same physical component.

2.3.7 Transformers and Gyrators

Figure 2.20 contains 2 possible causalities of transformers, as Bondgraph element at the left, equations in the middle and blocks at the right. The first couple of equations fixes output effort from input effort, and input flow from output flow, itself calculated elsewhere in the system. With the chosen power orientation, output power equals input power and consequently the same constant K appears. So the top illustrates one possible causality for transformers.

The next line of Fig. 2.20 displays the other possible causality of the transformer: f1 comes from the left and f2 leaves to the right; similarly e2 from the right determines e1 leaving to the left. One causality can be chosen freely, but then the other one is mandatory. So if the causal stroke on one bond is applied on the transformer side, on the other bond the stroke is necessarily away from the transformer. With a little experience, the reader immediately recognizes the correct causal strokes for transformers or otherwise, and of the following gyrator.

Figure 2.21 contains the causal stroke pattern, the equations and the blocks for a gyrator. For good impression, connections with effort in the blockdiagram are written above and flow connections below. For this, the connections have to cross since output effort (torque) is proportional to input flow (current in an electric motor). This is the reason why gyrators are also called overcrossed transformers.

Let us repeat that power conservation in transformers and gyrators enforces equality of the conversion parameters. There could be the objection that real transformers and electric motors do not conserve power but have a finite efficiency (90% for a good motor). The answer is that a really good motor is built from a gyrator for the essential action and other elements, particularly R-elements that dissipate power. Further, C- and I-elements that store energy and therefore momentarily disturb the power balance; further, they can produce vibrations and shocks (sec. 6.3). These additional R-, C- and I-elements are often called parasites, opposed to the essential elements, since they eat up power and disturb just like undesirable insects.

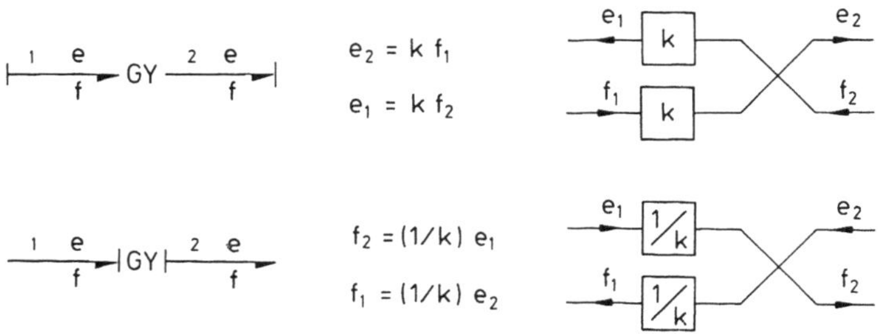

Figure 2.21 Two possible causalities for the gyrator, which is a kind of overcrossed transformer, since output effort e2 is proportional to input flow f1.

2.3.8 Causality Pattern and Power in Junctions

Figure 2.22 displays at the top a p-junction with usual power orientations. Power inflow on bond 1 and 3 is positive, as power outflow on bond 2. This imposes the sign for the flow equation as shown, since all efforts are equal. At the right the equations are represented as elements of block diagrams. Equality of efforts transforms into the branch points, and summing of flows in the mixing point, where the signs follow from the power orientations.

An s-junction with different power orientations appears in the middle of Fig. 2.22. Here we have a branch point with equality of flows and a mixing point for efforts with signs again controlled by power orientations.

The causal pattern of junctions is always the same:

- On a p-junction one effort pushes inward, all others outward.
- On an s-junction one flow points inwards, all others point outwards.

The bottom part of Fig. 2.22 contains a mnemonic for causal strokes of a bond.

- The end with the stroke looks like a piston and pushes effort in this direction
- the fine end of the bond looks like a needle, or like an injector, and points (or injects) flow in its direction.

It is good practice to visualize each bond equipped with a needle that points in this direction and a piston that pushes effort in the other. Imagination of this kind makes the scientist successful.

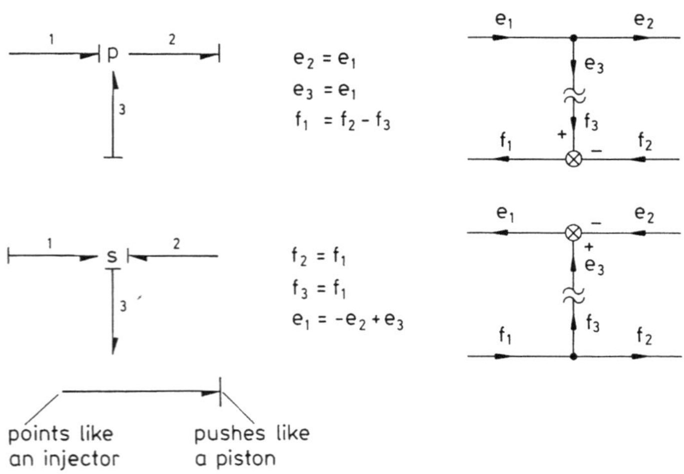

Figure 2.22 Causality pattern on junctions: A p-junction has one effort that pushes inwards, and all other efforts push outwards. Thus they become a branch point for efforts and a mixing point for flows. The s-junctions have the converse pattern. The bottom explains the terminology: The cross-stroke end looks like a piston that pushes (imposes) an effort to the right, and the fine end like a needle that points or injects a flow to the left.

Further, in going from a Bondgraph to a block diagram, a p-junction becomes a branch point for efforts and a mixing point for flows, an s-junction a mixing point for efforts and a branch point for flows. Naturally the above description remains true for junctions with four or more bonds. From this follows also the close relation between Bondgraphs and blockdiagrams:

A Bondgraph with all power orientations and causalities contains the same information as a blockdiagram, but much more compactly. Further it is easier to write and especially the insidious sign errors are easier to avoid.

2.3.9 Systematic Causality Assignment

Causality assignment can be made more efficient by the following procedure (Karnopp-Rosenberg 1975). It takes account of the fact that on choosing a free causality other causalities become mandatory: we have causal implications especially on junctions.

1. Select a source causality
2. Extend causal implications throughout the Bondgraph
3. If any source causality remains free goto 1, else goto 4
4. Assign a desirable causality
5. Extend causal implications
6. If any desirable causality remains free goto 4 else goto 7
7. Assign a free causality
8. Extend causal implications
9. If any causality remains free, goto 7, else write: causal assignment complete

If any extension above cannot be done, we have a causal conflict. Further, assigning free causalities leads to algebraic loops: for both we need the help in sec. 6.5.

2.4 Simulation from Blockdiagrams and Bondgraphs

The main purpose of writing a Bondgraph is to simulate it. The method is to first write a Bondgraph from a circuit, such as in Fig. 2.23 for a simple example of an electric circuit or a hydraulic servomotor (left and right below). Powers are already assigned out from the flow source into the one-ports and through the transformer. The Bondgraph is labelled not consecutively but such that labels 1 to 9 refer to the left of the transformer and 11 to 13 to its right. This is a practical trick for easier remembering and would not be needed on mathematical grounds.

The next step is to assign causalities by the following procedures

1. Assign mandatory causalities, here only for the flow source.
2. Select a desirable causality, say integral C2. By this selection several causalities have become mandatory, namely on bond 3, 4 and 11. The last two are still free.

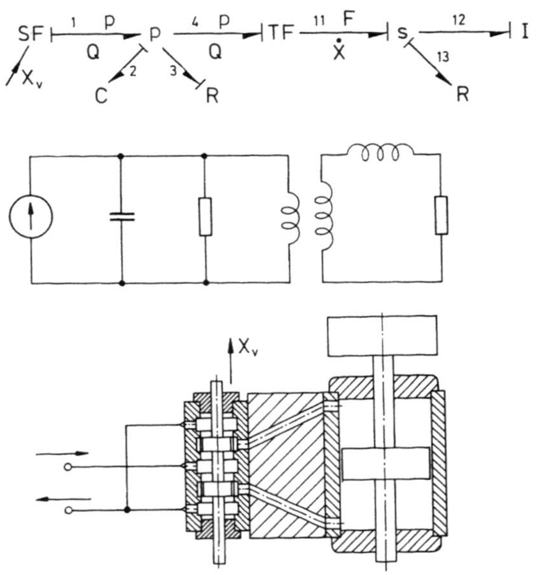

Figure 2.23 Simple Bondgraph with flow source, transformer and some oneports, corresponding to the electric circuit (bottom left) and to the hydraulic servomotor (bottom right).

3. Select another desirable causality, say I12. Thus the only remaining causality on R3 has become mandatory.

The result of the above assignment feature appears on Fig. 2.24 at the top as a Bondgraph, followed by the corresponding equations and on the bottom as a blockdiagram. In the equation some variables, namely Q2 and F12, have already

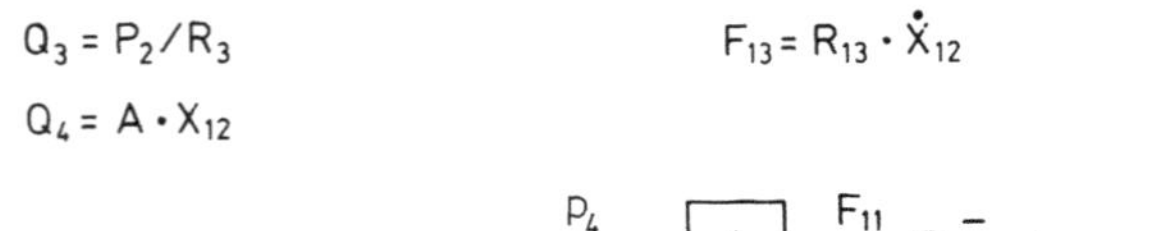

$$Q = k \cdot X_v$$

$$P_2 = \frac{1}{C}\int (Q_1 - Q_3 - Q_4)\,dt$$

$$Q_3 = P_2/R_3$$

$$Q_4 = A \cdot X_{12}$$

$$F_{11} = A \cdot P_2$$

$$\dot{X}_{12} = \frac{1}{I}\int (F_{11} - F_{13})\,dt$$

$$F_{13} = R_{13} \cdot \dot{X}_{12}$$

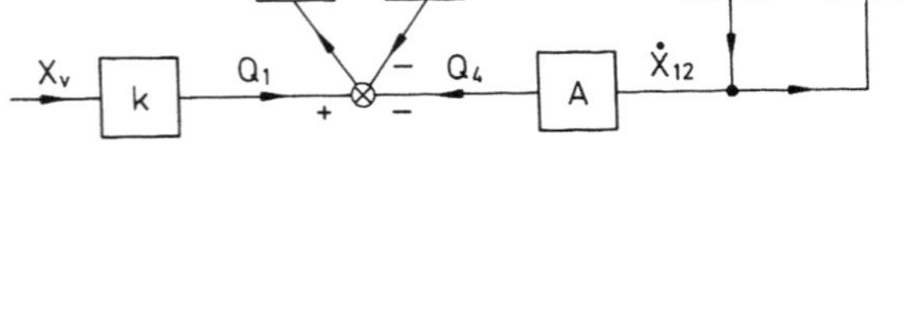

Figure 2.24 Same Bondgraph with assigned integral causalities followed by corresponding equations and the blockdiagram. Some intermediate variables, like Q2 and F12 have already been eliminated by reading over the junctions.

been eliminated by reading the equations of the junctions. The equations can then be programed with a suitable program, even in BASIC or FORTRAN. Instead of the general labels effort or flow we use p and F for efforts and Q and $\dot{x}$ for the flows corresponding to a hydraulic simulation to remind of the nature (dimensions) of the variables. Similarly the transformer parameter (sometimes called modulus) is now A, the area of the hydraulic cylinder. Further, the appearance of the blockdiagram with a block going forward in the upper line and returning below is typical for a diagram obtained from a Bondgraph.

The Bondgraph and blockdiagram of Fig. 2.24 are integral and good for computation. For human study it would be possible to arrange derivative causality on C2 and I12 without violating mandatory rules. This would lead to a simpler blockdiagram without the inner loops.

Figure 2.25 shows the listing and simulation result with the TUTSIM program using realistic parameters for a hydraulic servomotor, in practice a cylinder of 50 mm diameter and 1963 mm^2 area. The structure code at the right of

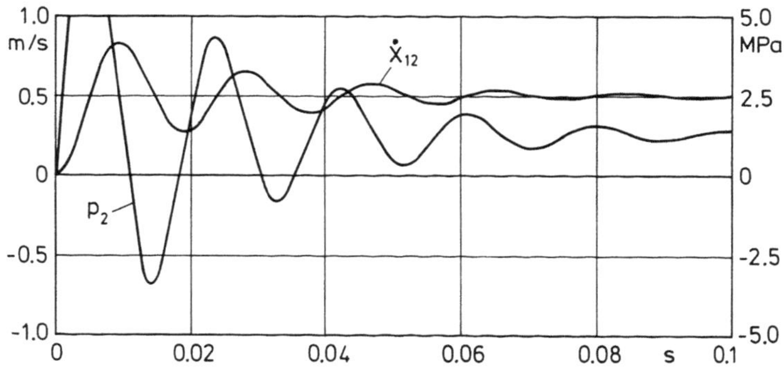


```
TIMING: 500.000E-06        0.1000000
PLOTBLOCKS AND RANGES
     0      0.0000          0.1000000
    12     -1.0000          1.0000
     2     -5.0000          5.0000

980.0000           1 SF      111                  ; servovalve flow
  0.3510000        2 C         1     -3     -4    ;fluid compressib
  0.0000
 11.6000           3 R         2                  ;leakage admittance
  1.963E+03        4 TF       12                  ;cylinder area
  1.963E+03       11 TF        2                  ;cylinder area
100.0000          12 I        11    -13           ;piston speed
  0.0000
  5.000E+03       13 R        12                  ;load friction
  1.0000         111 CON                          ;file bonsi24.sim
```

Figure 2.25 Listing and simulations result of Fig. 2.23 and 2.24 with the TUTSIM program. In the central part at the left is the structure code, almost the same as the equations in Fig. 2.24. Center left are the parameters for a hydraulic servomotor.

the list is almost identical to Fig. 2.24. Since we have here a simple second order system there is only one resonance that can still easily be calculated by hand (good student exercise).

We have simulated here from a Bondgraph with integral causalities from which we can write out the equations or transform into a blockdiagram. The blockdiagram is considerably more complex and contains only the same information as stated before. In case of systems of higher complexity it is better to avoid blockdiagrams and equations altogether and read the input code directly from the Bondgraph. This is possible for TUTSIM and other programs like ENPORT and CAMP.

The simulation procedure shown here on the simple example is also efficient for really complex systems. For nonlinear elements we need the block-Bondgraph of sec. 2.5.

The procedure is to write a Bondgraph without worrying about power and causality, and then to apply them as shown. The result is the augmented Bondgraph of Fig. 2.24, ready for computation. It also demonstrates clearly that on p-junctions only one effort can push and all but one flows point to the junction. On s-junctions one flow points and all minus one efforts push. Going to the blockdiagram we see how both junctions become branch and mixing points as explained before (sec. 2.3).

In this section we have demonstrated the systematic procedure to assign integral causalities and have taken the example to a simulation. Of course some undesirable derivative causality will come in more difficult cases, and also other troublesome features like algebraic loops; these and their remedies will be discussed in sec. 6.5.

2.5 Nonlinearities and Block-Bondgraphs

We have seen in sec. 2.4 that a fully augmented Bondgraph is equivalent to a block diagram. Here we shall introduce a further combination, where a general Bondgraph element is split into a limited one and into a block attached to one side of a bond. Such block-bondgraphs are especially useful for nonlinear C- and R-elements, as we shall now illustrate.

Take the example of a non-linear C-element, say a spring with stops with the F/x characteristics of Fig. 2.26 at the left. The C-element is really an integrator between $\dot{x}$ and x in the mechanical case, followed by a functional relationship, non-linear in general. This is shown as block diagram on the right of Fig. 2.26 using the TUTSIM symbol INT instead of 1/s for integration and FNC for the function.

The idea now is to combine the right hand side of Fig. 2.26 into a Bondgraph using the notion that effort and flow travel on the sides of each bond. This is accomplished on Fig. 2.26, bottom, by the block sitting on the top side of bond 23. Here the flow comes from the bond 24 over the s-junction, travels along the lower

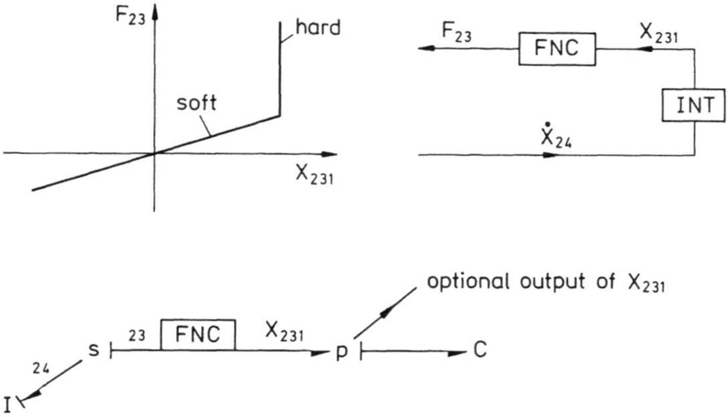

Figure 2.26 Non-linear spring characteristics (top left), block diagram with an integrator obtaining x231, followed by a function block FNC for the force with a soft and a hard part. Below representation as block-bondgraph where flow travels towards the C-element on the lower side without obstruction, but returning effort must pass the block on the top side and go through FNC. Optional output of x231 takes place through the p-junction.

side of bond 23 toward the C-element, where it is integrated. It is then labelled as x231 and travels backwards through the block on the top side to become F 23. Reasons for this particular labelling will be described in sec. 6.6.

The following remarks apply to Fig. 2.26.

1. The integrator INT of the block diagram (upper line) has been replaced by C in the Bondgraph which must be given a parameter of one to be a pure integrator; this maintains similarity with more general Bondgraphs.
2. The p-junction between the C-element and the block indicates the taking out of the position x231, which may be needed elsewhere for simulation (example Fig. 4.19).
3. In order to clarify whether the block modifies effort of flow we need the convention already mentioned, that flow travels along the bond side with the power arrow.

A worked example of a simple series resonator with a non-linear C-element appears in Fig. 2.27 as block-bondgraph; mechanically it would be an oscillating body with mass I14 on a spring with variable stiffness and damped by the (linear) resistor R12. The following Fig. 2.28 contains the TUTSIM listing where the FNC block 13 produces a force/displacement characteristic similar to Fig. 2.26 with a soft and a hard part. The simulation result on Fig. 2.28 shows the force peaks when the body reaches the hard part and consequent rapid change of speed.

Figure 2.29 contains a further example with an R-element in conductance causality, modulated by signal called x231. At the top there is a standard representation with a modulating signal bond, below the computable implementation using block-bondgraphs with the MUL block of TUTSIM and on the

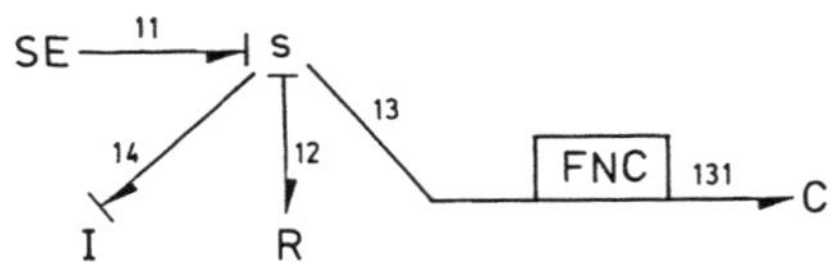

Figure 2.27 Block-bondgraph of an oscillator with a non-linear spring with similar characteristics.

bottom the equation, linear except for the multiplication with x231. Since the power arrow is on the same side as the block, we immediately see that flow is modified (multiplied) by the signal bond.

Block-bondgraphs as explained here are a powerful programing help for many practical simulations and we shall see later several more examples.

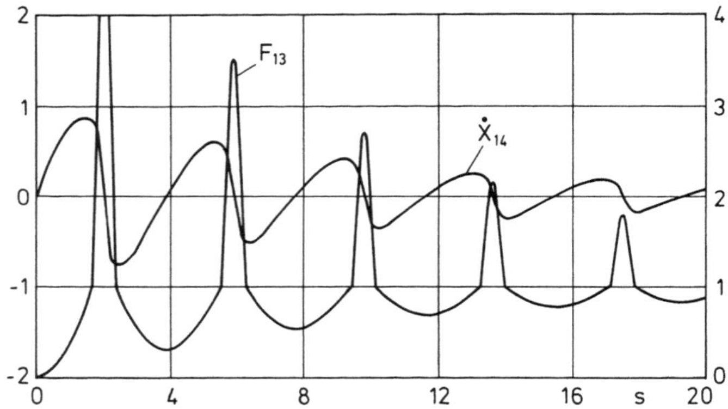


```
Timing:    0.0200000  ,DELTA  ;   20.0000      ,RANGE
PlotBlocks and Scales:
Format:
     BlockNo,  Plot-MINimum,  Plot-MAXimum;  Comment
Horz:     0 ,   0.0000     ,  20.0000     ;  Time
  Y1:    14 ,  -2.0000     ,   2.0000     ;
  Y2:    13 ,   0.0000     ,   4.0000     ;
  Y3:       ,              ,              ;
  Y4:       ,              ,              ;

   1.0000              11 SE      111                    ;force source
   0.2000000           12 R        14                    ;sliding friction
                       13 FNC     131
     1    -2.0000         -21.0000
     2    -1.0000          -1.0000
     3     1.0000           1.0000
     4     2.0000          21.0000
   1.0000              14 I        11     -12     -13
   0.0000
   1.0000             111 CON                            ;file bonsi25.sim
   1.0000             131 C        14                    ;basic spring
   0.0000
```

Figure 2.28 TUTSIM listing and simulation result of the non-linear oscillator. Upon entering the hard part of the spring characteristics the contact force F13 has sharp peaks.

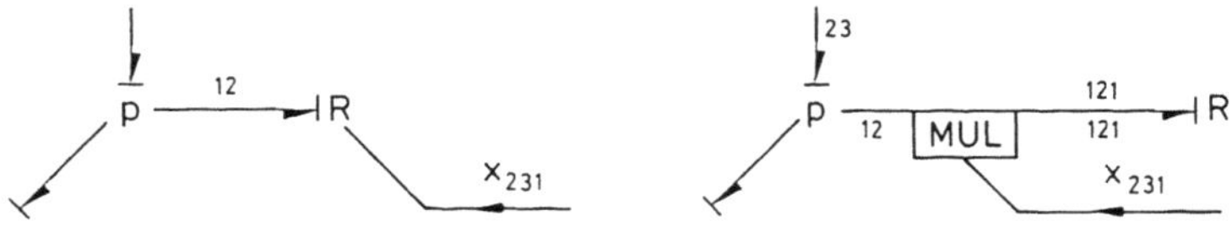

$$Q_{12} = \frac{1}{R_{121}} \times p_{23} \times x_{231}$$

Figure 2.29 Modulated resistor in conductance causality, at the top with signal bonds, below as block-bondgraph. The resistor is linear here in pressure but non-linear by multiplication with the position x231.

2.6 Multiport Bondgraph Elements

Although the Bondgraph elements presented so far allow the modelling of almost all systems that are suitable for Bondgraphing at all, it is convenient to combine some of them into multiport elements. Prominent are the C-, I- and R-fields, which in general are capacitor, inductor or resistor networks with two or more bonds. Convention has to call them "fields", but there is no relation to the word in the sense of field theory of distributed elements (sec. 1.2 or 8.2).

We would like to note here that multiport Bondgraph fields are another tool to represent complex systems concisely, yet also precisely for computation, as we shall exemplify below. Further they are powerful to explain some effects of physics more impressively than mere equations. The fields have the three levels of description mentioned before: Constitutive equation, energy or power relations and reversibility, each added with a small refinement as we shall show. We expect this since they are a natural extension of the one-port C-, I- and R-elements.

2.6.1 C-Fields

Figure 2.30 shows the symbol of a C-field or generalized capacitor network. Here the two efforts are functions of the two displacements, non-linear in general. They do not depend on the speeds $\dot{q}$ (flows $f = \dot{q}$), otherwise there would no longer be

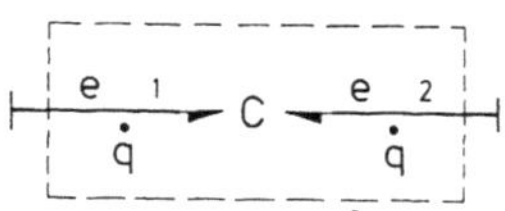

$$\underline{e} = \underline{\underline{K}}\,\underline{q} \qquad \begin{vmatrix} e_1 \\ e_2 \end{vmatrix} = \begin{vmatrix} K_{11} & K_{12} \\ K_{21} & K_{22} \end{vmatrix} \cdot \begin{vmatrix} q_1 \\ q_2 \end{vmatrix}$$

$$\underline{q} = \underline{\underline{C}}\,\underline{e} \qquad \begin{vmatrix} q_1 \\ q_2 \end{vmatrix} = \begin{vmatrix} C_{11} & C_{12} \\ C_{21} & C_{22} \end{vmatrix} \cdot \begin{vmatrix} e_1 \\ e_2 \end{vmatrix}$$

Figure 2.30 Bondgraph symbol and constitutive equation of a 2-port C-field. The equation represents two possible causalities and the stiffness matrix K is the inverse of the compliance matrix C. Both matrices are symmetric (Maxwell symmetry).

a C-field, but a combination of C- and R-fields. This happens with real capacitors that have non negligible leakage.

In the linear case the relation can be expressed as the matrix equation in the center of Fig. 2.30 with the stiffness matrix K. It can be inverted to become the compliance matrix C with its equation on bottom of Fig. 2.30.

The essence of C-fields is that the efforts are a function of both displacements, the coupling expressed by the coefficients K12 and K21. These are equal by famous Maxwell symmetry as derived in the box. Essentially the power entering through both bonds is saved as stored energy (called U in thermodynamics). Returning to the initial state (initial values of both displacements), the same energy leaves through the bonds. One would expect that this energy conservation restricts the constitutive equation, the relation between displacements and efforts, and this restriction is precisely the Maxwell symmetry. The stored energy function and the Maxwell symmetry is derived in Box 2.1, Page 38.

In the case of non-linear C-fields one can always make a series expansion around a point of operation and obtain Maxwell symmetry for the incremental quantities. Naturally in this case the parameters in the matrices depend on the point of operation.

2.6.2 Physical Meaning and Examples of C-Fields

Our first example is the electric capacitor network of Fig. 2.31. We find here two capacitors between the ground and wires 1 and 2 and further a coupling capacitor between both wires. This coupling capacitor generates the coupling terms in the constitutive equation; if it is small the coupling becomes negligible and the network behaves like two separate capacitors or separate C-elements.

The capacitor nework of Fig. 2.31 can also be represented by a Bondgraph with p- and s-junctions and 3 C-elements (Fig. 2.31, top right), but often it is more convenient to merge them in the two-port C-field below.

Figure 2.32 shows a mechanical example at the top left, a T-beam fixed rigidly to the floor. It is attacked by forces at points 1 and 2. We can now make some conceptual experiments to find the properties of C-fields:

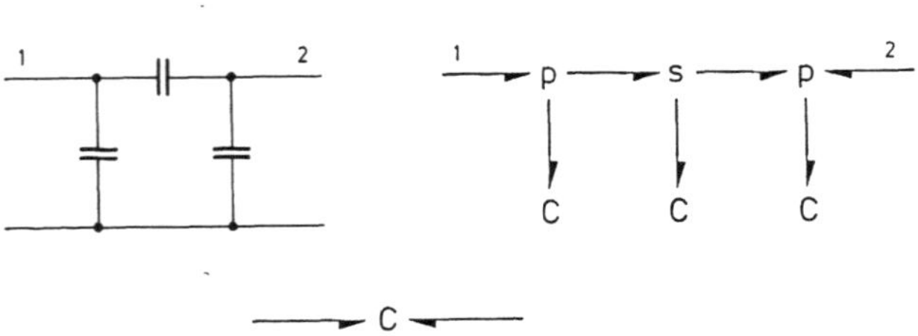

Figure 2.31 Electric 2-port C-field as capacitor network and corresponding C-elements joined by p- and s-junctions, below as Bondgraph symbol. Real capacitors with leakage have more complex behaviour.

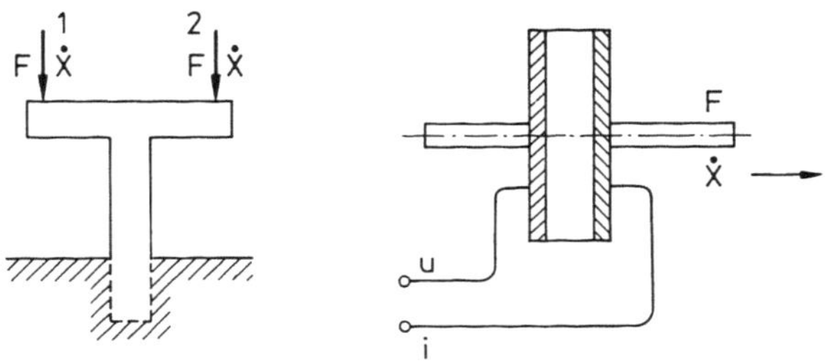

Figure 2.32 T-profile beam in a concrete floor as mechanical 2-port C-field and electric capacitor with moving plates as mechanical/electric C-field.

1. Applying variable force on point 1 and leaving point 2 free there will be a displacement of point 1 and also point 2 will move, so to speak it will wiggle under the variable load at the other point.
2. Again applying a variable force at point1, but preventing any movement of point 2, f.i. by a block and suitable screws or bolts, point 2 will apply a force on the block. Further point 1 will appear stiffer (have less compliance).

How much the effects are going to be depends on the relative stiffness of the flange and the central part of the beam, or mathematically on the constitutive equation.

What happens if we deform the T-beam so much that it reaches plastic deformation and remains deformed on load release? Well, then the beam no longer behaves as a C-field nor is it energy conserving or reversible.

The right part of Fig. 2.32 shows an electric capacitor with variable separation of the plates, a so-called moving plate capacitor. This is a mixed electromechanical Bondgraph element, the electric bond having voltage and current and the mechanical bond attraction force and speed. As calculated in the Box 2.2, the constitutive equation of the moving plate capacitor is simple but non-linear. The incremental quantities (efforts) show the required Maxwell symmetry.

Presuming a little on chapter 7, we should note that matter, especially the ideal gas of thermodynamics, also behaves like a C-field with two ports. Efforts are pressure and temperature, and flows are volume change and entropy current (entropy flow), definitely not heat flow as we shall explain.

2.6.3 TUTSIM Implementation of Two-Port C-Fields

The constitutive equations of linear C-fields can be programed in TUTSIM with the help of Fig. 2.33. We have on the bottom of the block diagram the 2 integrating blocks with inputs f25 and f23, calculating the displacements q293 and q313. The numbering is chosen to adapt easily to simulation as part of a large Bondgraph.

The stiffness matrix is represented by the gain blocks of Fig. 2.33, where no boxes (frames) are drawn around the designation GAI for economy of drafting. The central GAI is transversed by 2 signal connections. In principle there should

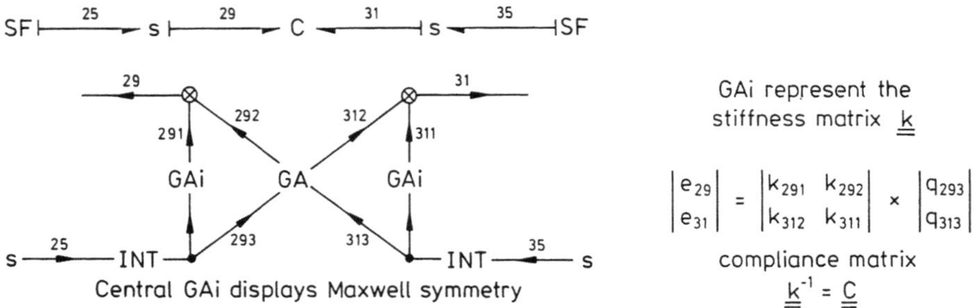

Figure 2.33 TUTSIM implementation of a 2-port C-field which is represented as the blockdiagram below (the frames or boxes have been omitted for economy). The central GAI block is run through by both crosscoupling connections.

be 2 separate blocks, but the combination usefully reminds us of Maxwell symmetry.

2.6.4 Reversibility and Power/Energy Relations in C-Fields

C-fields are reversible and energy conserving just as C-elements. So momentarily power can flow in through the bonds but will be stored as internal energy. It changes the state of the C-fields and must be evacuated through one or the other bond to return to the original state.

There are two remarks or refinements for C-fields with two or more ports:

1. The stored energy can leave the C-field through a different bond than it has entered. With the moving plate capacitor it can enter through the mechanical bond and leave through the electrical one. Running such a device through many cycles of loading and unloading we obtain an energy (power) converter from mechanical to electrical power.
2. Energy can enter as mechanical energy and leave as electrical energy as we have just seen. So the stored energy U has lost its (mechanical or electrical) character, it is just universal energy. In other words, there is no conservation of mechanical energy alone nor electric energy alone, rather energy can escape in another form. This is also the principle of heat engines as we shall see in chapter 7.

2.6.5 I-Fields or Generalized Inductor Networks

Figure 2.34 shows the 2-port I-field as Bondgraph symbol at the top. It is similar to the C-field with the flow and momentum instead of effort and displacement, respectively. Like C-fields it can have linear or non-linear constitutive equations, at least in principle. In practice, inertia actions are always linear as long as the velocities are small compared to the velocities of light where one would have to

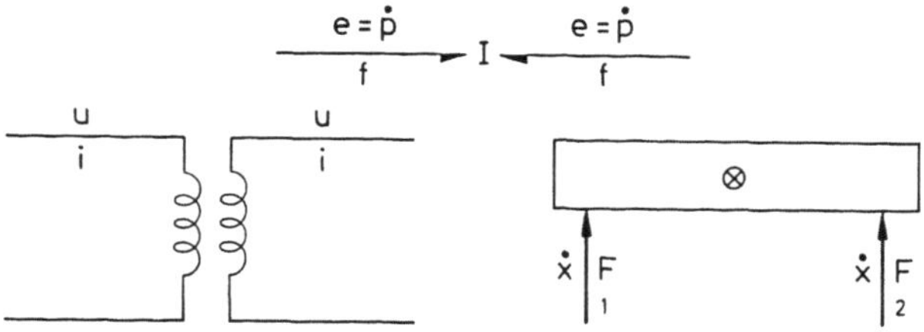

Figure 2.34 2-port I-field with example of coupled electric inductors and with an indeformable beam subject to mass forces.

use Einstein relativity. Further I-fields have a stored energy which is a function of the momenta and consequently they have Maxwell symmetry. Finally they are reversible.

The lower line of Fig. 2.34 contains examples of I-fields: Left 2 coupled inductors, which are non-linear in case of magnetic saturation, right an indeformable body with two points of application of force and with important mass action.

The difference between the bodies of Figs. 2.32 and 2.34 is as follows: In Fig. 2.32 the force application is so slow that inertia force can be neglected and deformation is important. In Fig. 2.34 the body is supposed so stiff that its deformation is negligible under the applied force. This is possible since the body is unrestrained and can move and accelerate.

2.6.6 R-Fields or Coupled Resistor Networks

Figure 2.35 displays the last of the Bondgraph multiports, a two-port R-field. Top left is the Bondgraph symbol, top right an example of an electrical network. If the coupling resistor is missing or very large we obtain 2 separate R-elements (as with the network of Fig. 2.31). R-fields establish relations between efforts and flows and can be linear or not.

Bottom left of Fig. 2.35 contains the constitutive equations of a linear R-field in resistance causality. Also here the resistors R12 and R21 are equal by the so-called Onsager symmetry. Such symmetry does not follow from simple principles such as the Maxwell symmetry, and was controversial for many years but can

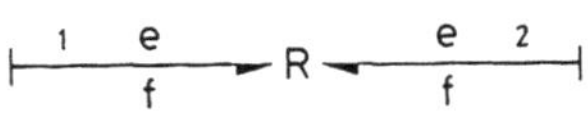

$$\begin{vmatrix} e_1 \\ e_2 \end{vmatrix} = \begin{vmatrix} R_{11} & R_{12} \\ R_{21} & R_{22} \end{vmatrix} \times \begin{vmatrix} f_1 \\ f_2 \end{vmatrix}$$

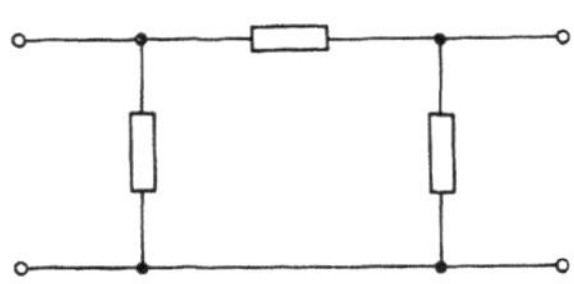

Figure 2.35 2-port R-field with electric example and constitutive equations. Also here the matrix is symmetric (Onsager's symmetry).

presumably be proven from difficult arguments of statistical mechanics.[1] Fortunately, practical Bondgraphers do not need to know them.

R-fields are power absorbing and irreversible in the sense of thermodynamics. The dissipated power is always positive and appears as general heat flow. Yet there is a little difference: If power is absorbed by the R-field on one bond, some power can emerge in the other bond. This is the principle of thermoelectric converters. Hence, the irreversibility applies only to the global dissipation, which must always be positive, but not to the single bonds.

2.6.7 Symmetry of Multiport Fields

Multiport fields can be built from one-ports connected by junctions, as was shown in Figs. 2.31 and 2.35 at right and will be again in sec. 5.2 with magnetic actuators. Such "build-up" or implicite multiports will always have Maxwell or Onsager symmetry. Only if gyrators are included the symmetries will disappear (Karnop-Rosenberg 1975).

Box 2.1 Stored energy functions and Maxwell symmetry in C-fields.

We place a conceptual control area around the C-field, indicated by the broken line on Fig. 1 and express the accumulation of energy inside:

$$\frac{dU}{dt} = e_1\dot{q}_1 + e_1\dot{q}_1$$

multiply by dt and manipulate:

$$dU = e_1\dot{q}_1 dt + e_2\dot{q}_2 dt$$
$$dU = e_1 dq_1 + e_2 dq_2$$
$$\text{Integrating: } U = U(q_1, q_2)$$

Effort can be obtained from the stored energy by derivation:

$$e_1 = \frac{\partial U}{\partial q_1} \qquad e_2 = \frac{\partial U}{\partial q_2}$$

With a further derivation we obtain from the equality of cross derivatives

$$\frac{\partial e_1}{\partial q_2} = \frac{\partial e_2}{\partial q_1} \text{ the important Maxwell}$$

symmetry from the equality of the cross derivatives

[1] The arguments seem not so difficult in the interesting little book of H.W. Kammer and K. Schwabe, Thermodynamik irreversibler Vorgaenge, Physik Verlag, Weinheim, Germany, 1985

Box 2.2 Equations for moving plate electric capacitor.

From basic electronics we have for the capacity

$$C = \frac{1}{C}q \qquad C = \frac{\varepsilon A}{x} \qquad \varepsilon = \text{Permeability}$$

$$U = \int dq\, u(x,q) = \frac{q^2}{2C} = \frac{q^2 x}{2\varepsilon A}$$

Attraction force at constant charge comes from derivation of the stored energy U with q = const

$$F = \frac{\partial U}{\partial x} = \frac{q^2}{2\varepsilon A}$$

The constitutive equation is simple, but not linear

$$u = \frac{xq}{\varepsilon A}$$

The incremental quantities become

$$du = \frac{x}{\varepsilon A}dq + \frac{q}{\varepsilon A}dx$$

$$dF = \frac{q}{\varepsilon A}dq + O\,dx$$

with the Maxwell symmetry of all C-fields

2.7 Junction Structures and Transformations

We will describe here some transformations of junction structures which help to simplify many practical Bondgraphs.

Figure 2.36 contains at the top 2 s-junctions, one behind the other, which can be combined into one s-junction as shown on the top right. The same combination applies to the 2 p-junctions in the lower line.

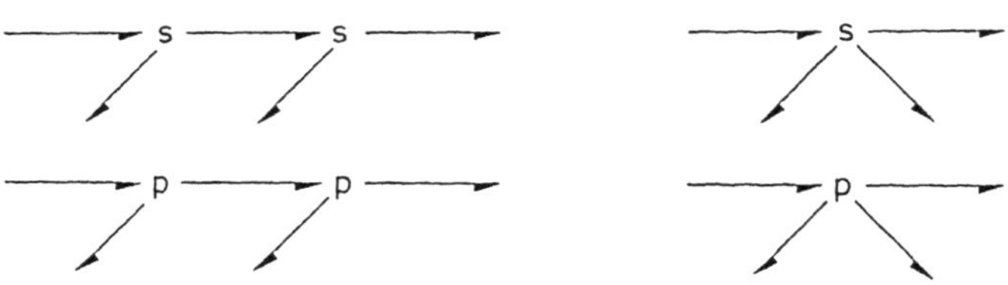

Figure 2.36 Possible contraction of two consecutive s- and p-junctions. Junctions with only two bonds are identities if power is oriented through.

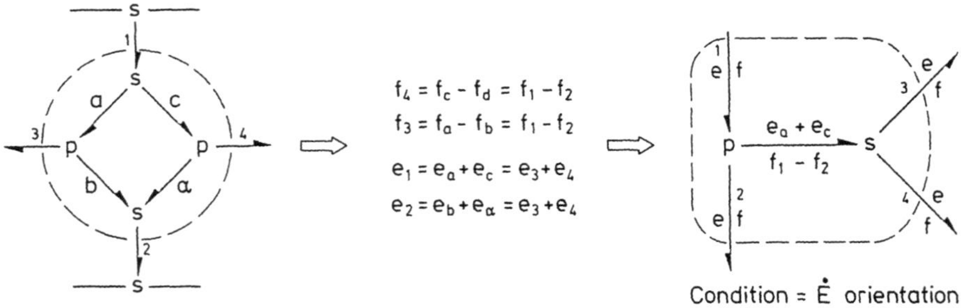

Figure 2.37 Diamond equality or transformation of Bondgraph structures for simplification. Such structures apply when writing mechanical Bondgraphs for moving reference frames.

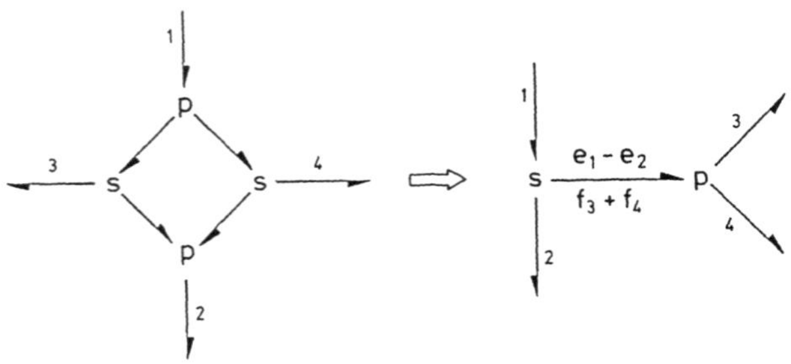

Figure 2.38 Another form of the diamond equality. The condition for both is the proper orientation of power.

Unfortunately, it is not admissible to interchange s- and p-junctions, but the following important diamond equality applies. Figure 2.37 contains at the top a junction structure (that looks like a diamond) which can be simplified into the lower structure. The proof is sketched by the equations on Fig. 2.37. However, there is a condition on the power orientation: equation: Power in bonds 3 and 4 must be the difference of powers in bonds a, b and c, d, respectively. Instead, there can also be the sums, but not the sum one side and on the other side the difference, in which case the equations would become singular.

Figure 2.38 contains another form of the diamond equality. All junction structures are power conserving and reversible. Such transformations are also the starting point for important algebraic theories of Bondgraphing, but for practical applications the main interest is to help simplify mechanical Bondgraphs. In fact, such diamond structures appear frequently with moving reference frames in mechanical systems (example in sec. 3.3) and similar problems in electricity.

3 Simulation and Design of Mechanical Engineering Systems

3.1 Opening of the Application Chapters

Good modelling and simulation has many faces and Fig. 3.1 gives an overview. We have started from the left and met Bondgraphs in the first 2 chapters. Now we are at a point where the way splits into the art of writing Bondgraphs (upper path) and into computation from Bondgraphs (lower path). Later they will approach each other again and become the Bondgraph way of life.

In principle, the following chapters 3 to 5 treat writing Bondgraphs and chapter 6 computation. In practice we shall give worked examples and thus alternate a little between the upper and lower path of Fig. 1.

The end is at the right of Fig. 3.1 called the Bondgraph way of life. This not only includes its simulation capability but another surprising benefit: With knowledge of Bondgraphs it is much easier to understand new machines and devices and to communicate or elaborate new ideas. Thus the Bondgraph becomes the vehicle for new ideas and inventions.

In the present chapter 3 we shall concentrate on mechanical systems which are simple in general but require some delicate points.

3.2 Mechanical Systems with Ground Reference

We start with the simple mechanical oscillator of Fig. 3.2. A body with mass can slide horizontally as driven by a force. It is connected to the fixed wall at the left, or the ground by a spring and a linear resistor.

The schematic and description implies that the Bondgraph should contain an SE (effort source), I-, R-, and C-elements (friction, mass and spring forces), but the question is to connect them correctly. Looking on the schematic of Fig. 3.2 we observe:

- All parts move with the same speed.
- There is a summing of the forces.

These ideas, expressed as equations on the bottom of Fig. 3.2 are exactly those of an s-junction. Then hence, the elements are connected by this junction as shown by the Bondgraph at the right, labelling here goes simply from 1 to 4.

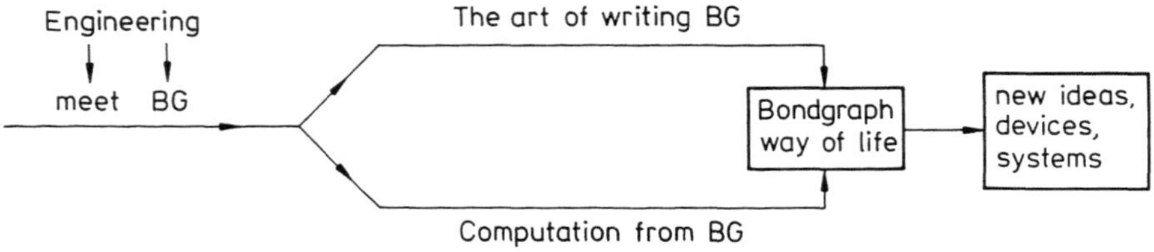

Figure 3.1 Overview of our study plan of Bondgraph modelling. Soon after the start our work splits into the art of writing Bondgraphs and into questions of computation. Later, Bondgraphs can explain new or unfamiliar engineering devices.

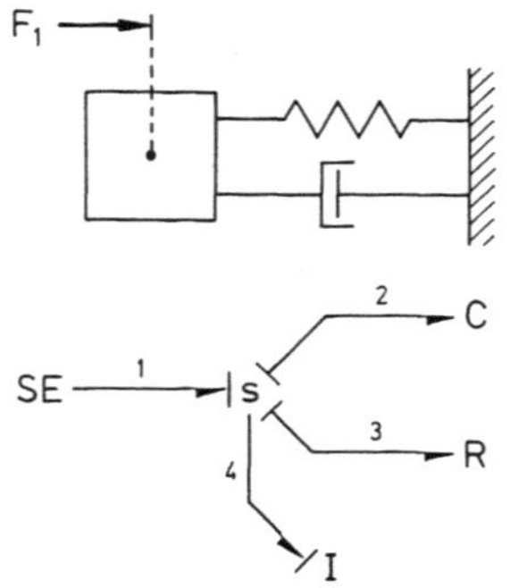

$$F_4 = F_1 - F_2 - F_3$$

$$\dot{X}_1 = \dot{X}_2 = \dot{X}_3 = \dot{X}_4$$

Figure 3.2 Mechanical oscillator with one degree of freedom: Scheme at left, Bondgraph at right and the equations of the junction at bottom.

We wrote the Bondgraph without worrying about power directions or causality. Now we assign power directions: Out from the source and into the one-ports (I-, R-, C-elements). The next step is systematic causality assignment:

Mandatory is only the effort of the effort source. Next we chose as desirable the integral causality on bond 4 (needle out of I-4), then the other causalities 2 and 3 are mandatory (efforts out of them), since an s-junction accepts only one needle (flow input).

Figure 3.3 contains at the top the equations obtained from the Bondgraph by inserting the constitutive equations of the elements. (We have set zero for the

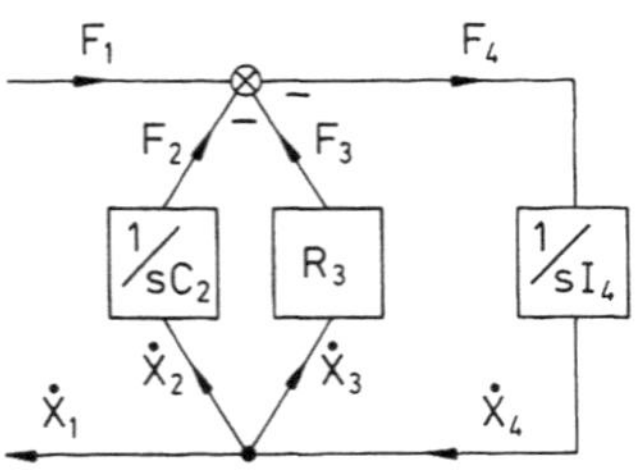

$$F_4 = F_1 - \frac{1}{C_2}\int \dot{X}\,dt - R_3\dot{X}$$

$$\dot{X} = \frac{1}{I_4}\int F_4\,dt$$

Figure 3.3 Complete equations of the simple oscillator, obtained from Fig. 3.2 by inserting the equations of C-, I-, and R-elements (top) and corresponding blockdiagram (bottom).

common speed.) The equations are integral as desired. In particular, the equation for $\dot{x}$ is Newton's acceleration law in integral causality. The lower part of Fig. 3.3 contains the corresponding blockdiagram with efforts (forces) in the top and flows (speeds) in the lower lines. It also illustrates how the s-junction splits into a mixing point (addition or substraction) for the efforts and a branch point for the flows.

The signs of the equations for efforts and the blockdiagram follow from the power orientations of the Bondgraph. In words we could say that acceleration force (F4) equals input force (F1) minus friction and spring forces (F2, F3).

One question would be: What happens if the wall in Fig. 3.3 starts to move under the spring and friction forces? Well then we need a Bondgraph for moving reference points as treated in sec. 3.3.

3.2.1 Motor and Pump Drive

Our next example is a motor, presumably electric and generally of any type, which turns at a constant frequency of rotation and drives a pump through a gear reducer. Assume that the motor turns at a constant 150 Rd/s (1440 R.p.m.) and the pump needs 50 Rd/s, we need a reduction factor of three.

Figure 3.4 contains no schematic, which would be difficult to draw without pictures of the actual components, but a word Bondgraph on top with power orientations. Causalities can be fixed only when the words are replaced by the standard elements of chapter 2 because only they impose desirable or mandatory causalities.

The lower part of Fig. 3.4 contains a Bondgraph with standard elements, except for the pump which is left unspecified. Integral causalities have been applied, power directions fixed; labels have been selected such that between motor and gear reducer labels 11 to 19 apply and between gear reducer and pump the labels 21 to 29. This is much better than consecutive labels and we need a few empty numbers if we want to later apply more detailed effects on either side of the gear reducer.

The Bondgraph Fig. 3.4 contains the following ideas:

- The motor turns at a constant rotation frequency because of SF11.
- The shaft is supposed to have rotary compliance (spring action) and a slip, leading to C12 and R13.
- C12 and R13 experience the same torque and the rotation frequencies are deducted from the motor rotation frequency. Hence, they are connected by p-junctions.
- The gear reducer is TF19 21 with its moment of inertia I24 and bearing friction R23. These elements experience the same rotation frequency but require torque; hence, they are connected through the s-junction.
- Nothing is fixed yet about the pump, but causality indicates that we need torque as a function of speed for simulation.

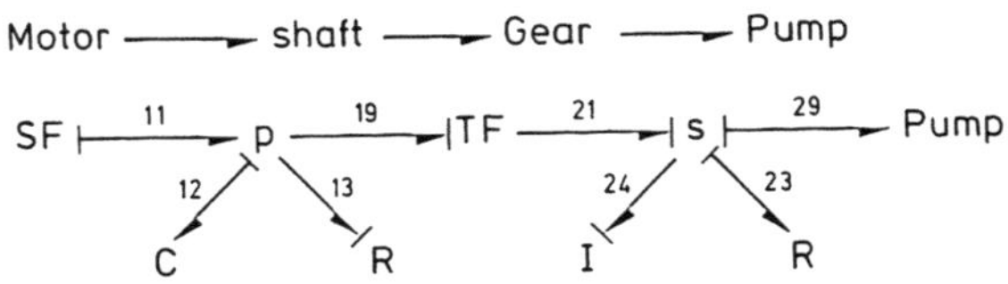

Figure 3.4 Bondgraphs of a motor and pump drive: Top as word Bondgraph and bottom as Bondgraph with standard elements.

Let us distinguish the role of the two R-elements in Fig. 3.4: R13 gives a slip, but no torque loss, R23 gives no slip but transmits torque to the ground (bearing friction between shaft and ground). These resistors lead to the slip and torque efficiencies, respectively, as we shall show later (sec. 3.5).

The equations between the variables of Fig. 3.4 are easy to write out, similarly as done in Figs. 3.2 and 3.3. We shall not give them here, nor a blockdiagram, which would already be complex. It is sufficient to note that the C-elements can interact through the transformer with the I-element to give torsional oscillations which are damped by both R-elements.

3.3 Systematic Writing of Mechanical Bondgraphs with Moving Reference Points

For applications such as (electric) motors with rotary casing, coupled railway wagons or planetary gear reducers we need a more systematic way to write Bondgraphs. It should help to incorporate the moving reference points or points of attack of the one-port elements. Before developing it by examples we shall describe the systematic procedure in a few words.

1. Assign an s-junction to all points with a definite velocity or frequency of rotation.
2. Put all elements, except I-elements, between the appropriate s-junctions behind a p-junction.
3. I-elements are directly attached to the s-junction representing the speed (or rotation frequency) of the appropriate point or machine element.

In fact, I-elements have no reaction force (they are attached to inertial space) but all other elements have. This follows from momentum conservation (Newton's 2nd law) and from the fact that I-elements are stores of momentum.

The resulting Bondgraph can be simplified by grounding one s-junction; this means to choose a frame of reference where the speed of this point is zero.

We shall now give some examples:

3.3.1 Coupled Railway Wagons

Instead of the fixed wall of Fig. 3.5, let there be another body or wagon, followed by a third one. The resistors and springs now represent the couplings between the

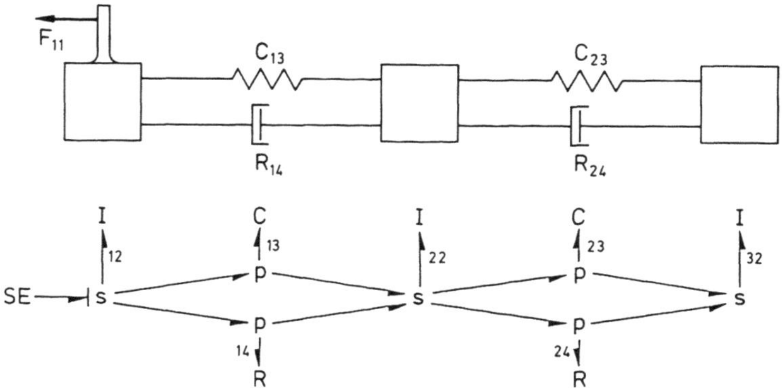

Figure 3.5 Coupled railway wagons, schematic at the top and systematic Bondgraph below. Each car has a definite speed as represented by an s-juntion. Further it has couplings, springs and resistances acting between the s-junctions through p-junctions. Mass forces and the locomotive driving force act only on the respective s-junctions.

wagons and we are concerned about longitudinal oscillations. For systematic writing we proceed as follows:

Each car runs with a definite velocity and is given an s-junction. The acceleration forces act on each car according to its mass. This is properly represented by an I-element connected to the corresponding s-junction. Similarly, the driving force acts on the car at the left (locomotive) represented by the effort source at the left.

The coupling R- and C-elements of Fig. 3.5 work between the cars: They experience the difference of car velocities and apply equal forces to both those cars. Hence, they are connected by p-junctions (equal forces, summing of flows) to the s-junctions representing the speeds of the adjacent cars. This leads to the Bondgraph of Fig. 3.5. Here we have already assigned the normal power directions: Out from the sources and into the one-ports. On the p-junctions it is necessary to have the power orientations as shown, since the coupling C- and R-elements experience the difference of the car speeds.

Figure 3.6 contains the Bondgraph for the coupled wagons after simplification from Fig. 3.5 by the diamond theorem (sec. 2.7). The causalities have been assigned all integral without difficulty. Labels have been added for most bonds, such that each decade (of numbers) refers to one car and to the coupling on its right side. It is possible to add more cars, say with bonds 42 to 82 for useful simulation of longitudinal oscillations of long trains.

Another modification of Fig. 3.6 would be to restrain (block) car 32 from movement or equivalently to give to I32 a very large (infinite) mass. In both cases $\dot{x}32$ becomes zero, so bond 32 disappears (it has no power). Then the nearby p-junction has only 2 bonds, is an identity and can be left out. Next the s-junction between C23 and R24 can be combined with I22. So I22 will be tied to the ground

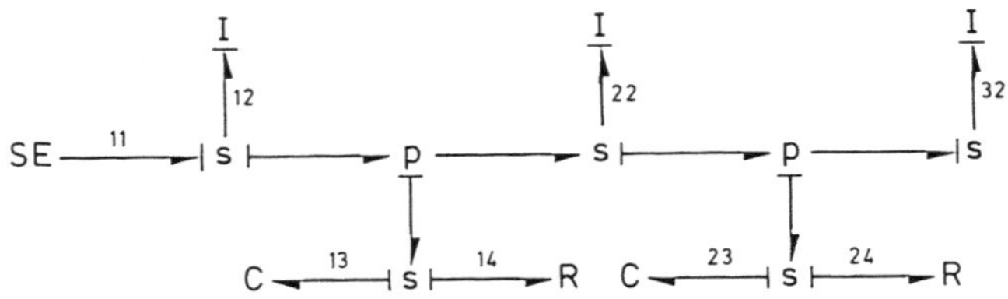

Figure 3.6 Bondgraph of the railway train simplified from Fig. 3.5 by the Diamond theorem.

by C23 and R24 similarly as in Fig. 3.5. Consequently, our Bondgraph represents 2 coupled oscillators, one of which is connected to the ground.

3.3.2 Reaction Torque of Electric Dynamometers and Gear Reducers

Electric motors supply torque through their shaft and produce an equal reaction torque in their casing that generally goes through the mountings to the ground. The reaction torque is a consequence of the principle of conservation of angular momentum. In Fig. 3.7 at the top there is a conceptual control area around the motor and the angular momentum supplied to the shaft must come from somewhere, namely the ground.

Similarly Fig. 3.7 at the top right represents a gear reducer; here the output torque or flow of angular momentum is supplied in part by the input shaft and the remainder by the mounting.

The lower part of Fig. 3.7 contains the Bondgraphs written according to our systematic way: Each point with a definite rotation frequency is represented by an s-junction. Here we have the shaft of the electric motor and shafts a and b of the

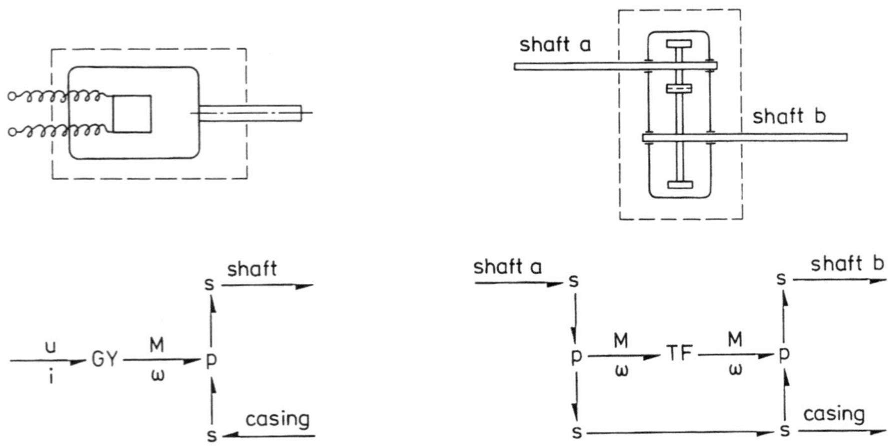

Figure 3.7 Electric motor (left) and gear reducer (right) with systematically written Bondgraph below. Broken lines indicate a control area for inflow and outflow of angular momentum.

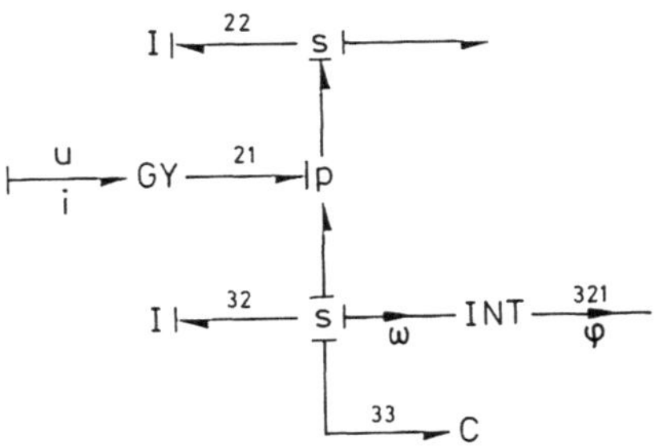

Figure 3.8 Electric motor with rotatable casing and fixed by a spring (C33) to the ground. Shaft inertia I22 and casing inertia I32 are added. This structure acts also as a dynamometer (torquemeter) where the angle φ 321 is proportional to torque.

gear reducer and the casings of both. The essential action of the electric motor is a gyrator that acts through the p-junction between shaft and ground, that is between both s-junctions. This expresses the fact that electromagnetic torque acts both on shaft and casing and the rotation frequency seen by the electric parts etc, equals the difference of shaft and casing frequency.

Similarly, the gear reducer acts as a transformer on 2 p-junctions between the shafts and the casing, each represented by an s-junction.

The electric motor can become a dynamometer by measuring the reaction torque, f.i. by connecting the casing to the ground via a spring and measuring its displacement as shown in Fig. 3.8. Here C33 is the spring and we have added I22 and I32 for the moments of inertia of shaft and casing respectively. Causality and power directions were assigned without difficulty. The integrator element INT 321 calculates the angle which is proportional to torque in principle. In practice the Bondgraph of Fig. 3.8 represents an oscillatory system which needs to be damped by suitable resistances, f.i. for bearing friction between shaft and casing. Of course in modern practice C33 is replaced by an electric load cell which is a very stiff spring element.

For an accurate dynamometer there are 2 sources of error in Fig. 3.8.

1. The electric wires should be such that they exert no torque on the casing. This can be achieved in spite of substantial wire tension by arranging the wires to be on a radial line from the shaft.
2. More serious is that many electric motors have cooling fans that carry away substantial angular momentum by the rotary movement of the air. The solution is to eliminate the rotary air by suitable baffles or to arrange airflow radially away from the axis of the casing.

These considerations demonstrate that there are many additional problems when realizing a precision instrument instead of a standard machine.

3.3.3 Planetary Gear Reducer

We obtain the Bondgraph for a planetary gear reducer from Fig. 3.7 by allowing the casing to rotate. In mechanical design, this would be the rotating cage of some

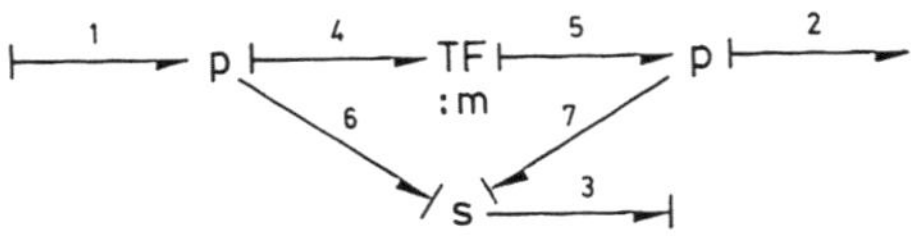

$$\omega_5 = m \cdot \omega_4 \quad \text{(a)}$$
$$\omega_2 = m \times \omega_1 - (1 + m)\,\omega_3 \quad \text{(b)}$$
$$M_1 = m\,M_2 \quad \text{(c)}$$
$$M_3 = -(1 + m)\,M_2 \quad \text{(d)}$$

Figure 3.9 Bondgraph of a planetary gear reducer from Fig. 3.7 where the casing becomes the planet carrier. The equations below come from the transformer and junction equations for the flows, and from power conservation for the efforts (torques).

gears, called planets; this is because they spin around their own axis and orbit around a central gear, called sun gear, just as in astronomy the real planets spin and orbit around the sun. The rotating cage is often called planet carrier.

Figure 3.9 contains at the top the Bondgraph with causalities and power directions and below the corresponding equations: Equation (a) is the frequency ratio of the shafts with the planet carrier stopped ($\omega 3 = 0$); equation (b) with planet carrier moving is determined from reading over the junctions. Equations (c) and (d) follow from power conservation.

The torque equations are a special case of the Karnopp power theorem (sec. 8.1) and can also be obtained by matrix multiplication. Here we observe 2 different points:

- The planetary gear accepts two flow inputs and has thus two degrees of freedom. So $\omega 1$ and $\omega 3$ can be chosen at will but then $\omega 2$ is fixed.
- The gear accepts only one torque input, M2 as assigned, but then M1 and M3 are fixed, that is determined by the mechanism.

Of course other causality assignments are possible for this mechanism and result in different but equivalent equations.

Our planet gear reducer is a special case of a mechanism (which are generally built with MTF, sec. 8.1) and demonstrates another theorem of kinematics: If a mechanism accepts n causal flow inputs, then it has n degrees of freedom. Here we had $n = 2$.

An automobile differential is a variant of a planetary gear reducer as shown in Fig. 3.10 with different causalities. It could be obtained from the preceding figure by Bondgraph transformations (TF slipping and combining of both p-junctions, sec. 6.4) but it is quicker to write the equations directly as in Fig. 3.10 at the right. The shaft of bond 1 comes from the planet carrier, bond 3 in Fig. 3.9 and corresponds to the cardan shaft in rear wheel driven automobiles. We see from the p-junction of Fig. 3.10 that there are equal torques in the wheels (bonds 2 and 3). In automobiles this leads easily to spinning wheels without traction if one is on ice. To remedy such wheel spin, limited slip differentials have been developed,

---

We know that transformers and junctions conserve power, hence, the whole structure of Fig. 3.9 is power conserving. Equations are written by first setting $\omega 3 = 0$ and secondly $\omega 1 = 0$.

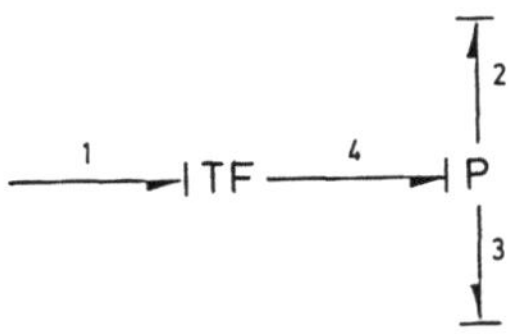

$\omega_1 = m\,\omega_4$

$\omega_1 = m\,(\omega_2 + \omega_4)$

$M_4 = m\,\omega_1$

$M_2 = M_3 = m\,M_1$

Figure 3.10 Bondgraph for an automobile differential, simplified from Fig. 3.9 and with different causalities.

which simply add controlled resistors to the Bondgraph of Fig. 3.10. They provide a slip coupling between shafts or bonds 1 and 2. A Bondgraph would be simple to write, but the technical realization is difficult, as always in real life.

3.3.4 Lanchester Damper for Torsional Oscillations

Mechanical drive systems with inertias and shaft compliances can make violent and dangerous torsional oscillations. They are damped by frictions or R-elements like bearing friction, but usually insufficiently. One way would be to increase the bearing friction, but this leads to unacceptable power loss and low efficiency.

In this situation, Mr. Lanchester came up with the idea (about 1905) to introduce additional friction connected to a flywheel instead of to the ground. So for steady rotation, the flywheel will take up the same rotation frequency as the shaft without power dissipation. With unsteady (AC or accelerating) rotation, there will be relative movement between flywheel and shaft, leading to dissipation in the coupling resistor. Originally, the Lanchester damper was used in aeroplanes between engines and propellers.

Figure 3.11 shows at the top a schematic and below a Bondgraph of the Lanchester damper on a shaft between inertias I12 and I22 and frictions R13 and R23. The rotary compliance of the shaft is divided into C141 and C142 and connected by p-junctions. The central part of the shaft is assumed to have a

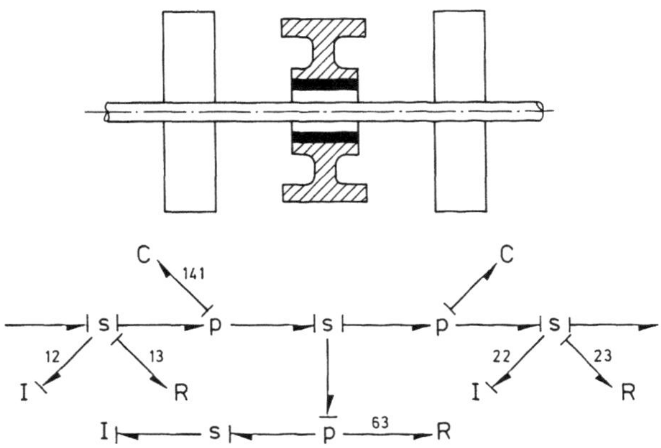

Figure 3.11 Lanchester damper with a flywheel able to rotate on a shaft and coupled by R63. Schematic on top, Bondgraph below. It works well for frequencies above R63/I62.

definite rotation frequency and is represented by an s-junction, also the flywheel itself (I22) by the s-junction below. The coupling resistor R63 is placed between both s-junctions behind a p-junction according to our standard procedure. Choice of power direction and causality is easily done.

Lanchester dampers are very efficient for frequencies above R63/I62, the reciprocal time constant of flywheel inertia and coupling friction; only in practice the implementation of a sufficiently high coupling resistor presents difficulties. The resistor does not have to be linear nor have the cylindrical form in Fig. 3.11. In a more recent application for computer printers (ca 1972), the flywheel is a disc pressed axially by permanent magnets to a disc fixed on the shaft; this leads to dry (Coulomb) friction.

The damping effect can be increased by a coupling spring between shaft and flywheel, but then becomes frequency sensitive, that is most efficient for the frequency defined by resonance between flywheel inertia and coupling spring. With violent torsional oscillations, the power dissipated in R63 becomes significant and cooling should not be forgotten.

3.4 Slip, Clutches and Electric Induction Motors

In mechanical engineering there are elements that introduce a controlled slip between two shafts without any significant torque loss. An example is the automobile clutch with a slip between engine and (manual) gear transmission; it is controlled by the position of the clutch pedal. The key point in writing a Bondgraph for the clutch is the slip without torque losses; hence it is represented by an R-element connected to both shafts by a p-junction as shown in Fig. 3.12. At the left there is a compact Bondgraph with a controlled R-element R24 and at the right a block Bondgraph with the R-element in resistance causality and a multiplication block. Here $x = 0$ corresponds to an open (disconnected) clutch and a large x leads to a tight coupling with minimum slip.

For numerical reasons, we cannot make R24 or X242 too large, because this would form a very short time constant $\tau = I23/R24$ (see sec. 6.3). Consequently the clutch reticulation of Fig. 3.12 will always have a small slip even when fully engaged. The clutch is then a controlled slip device and its efficiency equal to the speed ratio ($\omega 22/\omega 12$), the so-called slip efficiency. Of course some parasitic

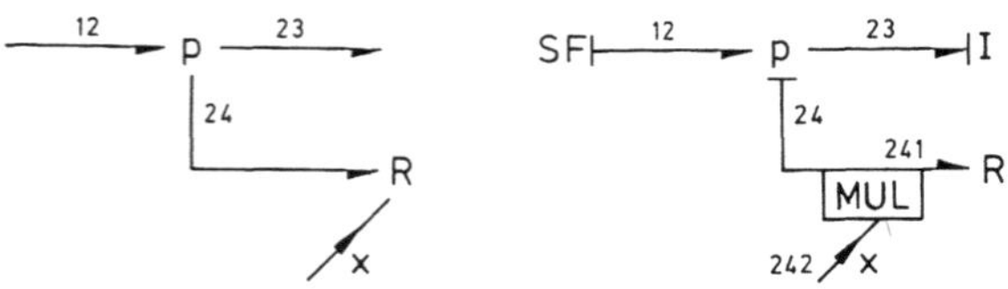

Figure 3.12 A clutch represented as a parallel resistor, modulated by a position (signal) x, as f.i. used in automobiles. Compact Bondgraph at the left and reticulation with block Bondgraph at the right with R241 in resistance causality.

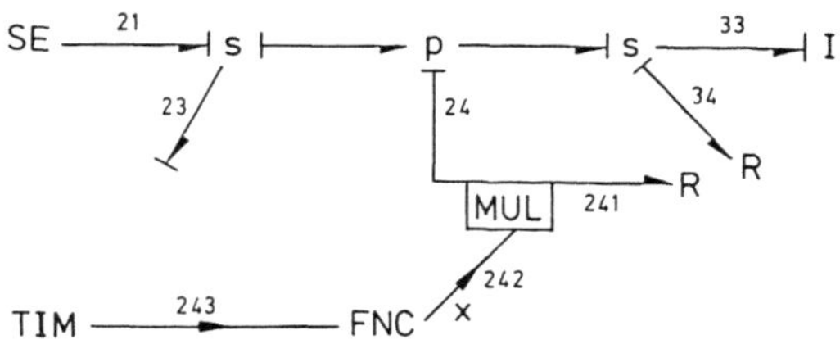

Figure 3.13 Reticulation of a clutch between an effort source SE21 with inertia I23 and a load represented by I33 and R34.

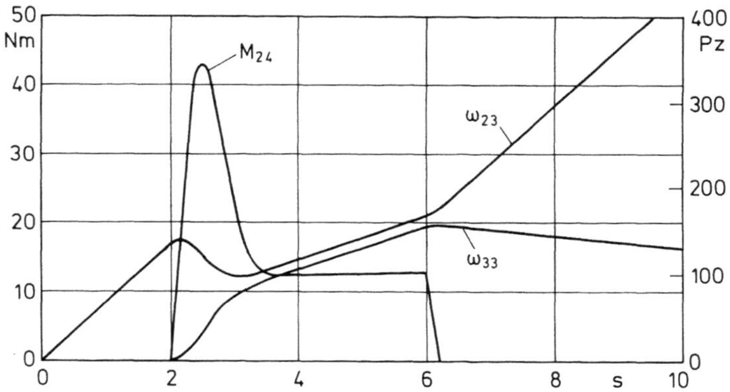


```
Timing:   0.0200000  ,DELTA  ;  10.2000     ,RANGE
PlotBlocks and Scales:
Format:
     BlockNo,  Plot-MINimum,   Plot-MAXimum;  Comment
Horz:      0 ,   0.0000     ,   10.0000     ; Time
  Y1:     24 ,   0.0000     ,   50.0000     ; coupling torque
  Y2:     23 ,   0.0000     ,  400.0000     ; moment o inertia
  Y3:     33 ,   0.0000     ,  400.0000     ; load inertia
  Y4:        ,              ,               ;

  20.0000          21 SE     211                ;torque source
   0.3000000       23 I       21     -24        ;moment o inertia
   0.0000
                   24 MUL    241     242        ;coupling torque
   0.4000000       33 I       24     -34        ;load inertia
   0.0000
   0.0200000       34 R       33                ;load friction
   1.0000         211 CON                       ;file bonsi34a.sim
   1.0000         241 R       23     -33        ;basic torque
                  242 FNC    243
     1     0.0000         0.0000
     2     2.0000         0.0000
     3     3.0000         1.0000
     4     6.0000         1.0000
     5     6.2000         0.0000
     6    10.0000         0.0000
                  243 TIM
```

Figure 3.14. Simulation of start up of the clutch of the preceding figure. The clutch is engaged gradually after 2 seconds and disengaged suddenly after 6 seconds. Shaft 23 slows down on engagement and shaft 33 speeds up. On disengagement shaft 23 accelerates more and shaft 33 slows down under friction R34. TUTSIM listing below.

elements, like bearing friction, will introduce small torque losses, but drag efficiency is usually very near to one (see sec. 3.5).

Figure 3.13 contains an example of an installation with a clutch, driven by the torque source SE21 and associated inertia I23. The clutch comprises the p-junction, the resistor R241 and the multiplicator MUL24 where the torque 241 is multiplied by a signal 242; it simulates the variable engagement of the clutch as generated by the function generator FNC242 and the time block TIM243.

The simulation result appears in Fig. 3.14 starting with an open clutch and gradual engagement between 2 and 3 secs; after 6 secs it is quickly disengaged. So shaft 23 speeds up at first, loses speed on engaging while shaft 33 speeds up and torque M24 is important. Then both shafts increase speed in parallel. On disengagement M24 falls to 0, shaft 23 speeds up more quickly and shaft 33 gradually loses speed under friction from R34.

Summarizing clutches, they always have equal input and output torques and their simulation works fine with the coupling resistor in resistance causality, even for the complex engagement cyclus of Fig. 3.14. Alternatively conductance causality could also be used but requires a finite rest torque for the disengaged clutch for numerical reasons. Which causality is preferable in the end depends on the whole system.

3.4.1 Electric Induction Motors

Electric induction motors of the asynchronous type are widely used because of low cost and high reliability; they operate similarly to a constant speed source followed by a fixed mechanical clutch. In fact, without torque, the shaft revolves exactly at the synchronous frequency (157 or 314 RAD/sec corresponding to 1500 or 3000 rpm at 50Hz AC). In order to produce a motoring torque a slip is required, about 50 rpm for rated torque leading to the usual 1450 rpm motors. Driving the motor above 1500 rpm (or the synchronous frequency) it acts as a brake and returns power to the electric wires.

Figure 3.15 contains the Bondgraph for such an AC motor with SF12 for the shaft turning at the synchronous frequency, I23 the rotor inertia and R25 for bearing and/or load friction. The slip is generated in R14 (more precisely in FNC14 and R141). In fact, the slip is linear near the synchronous frequency (points 1 to 3 of FNC14 in Fig. 3.16), but then varies according to the well-known

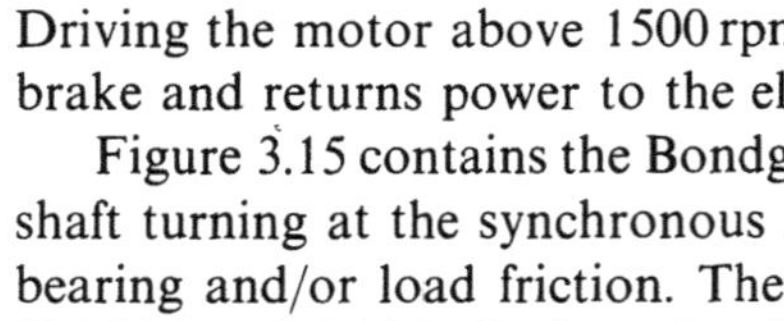

Figure 3.15 Reticulation of an AC induction electric motor. Resistor R14, that is R141 and FNC14 model the nonlinear slip/torque characteristic.

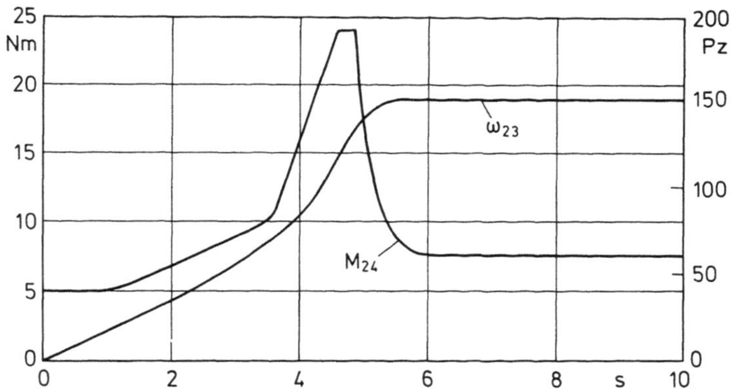


```
Timing:    0.0200000  ,DELTA  ;  10.2000      ,RANGE
PlotBlocks and Scales:
Format:
     BlockNo,  Plot-MINimum,   Plot-MAXimum;  Comment
Horz:      0 ,   0.0000      ,  10.0000     ; Time
  Y1:     14 ,   0.0000      ,  25.0000     ;
  Y2:     23 ,   0.0000      , 200.0000     ; motor mome o inertia
  Y3:        ,               ,              ;
  Y4:        ,               ,              ;

 157.0000           12 SF      121                ;driving rota frequency
                    14 FNC     141
     1   -7.5000        -10.0000
     2    0.0000          0.0000
     3    7.5000         10.0000
     4   22.5000         24.0000
     5   40.0000         24.0000
     6   60.0000         20.0000
     7   90.0000         10.0000
     8  140.0000          5.0000
     9  157.0000          5.0000
   0.2670000        23 I        14     -25        ;motor mome o inertia
   0.0000
   0.0500000        25 R        23                ;bearing/fan friction
   1.0000          121 CON                        ;file bonsi34b.sim
   1.0000          141 R        12     -23        ;basic friction
```

Figure 3.16 Start up of the AC induction motor of the preceding figure. The torque goes through a maximum at 5 seconds, leading to quick acceleration. Later shaft speed $\omega 23$ remains constant and torque M14 drops to a value determined by R25.

torque slip relationship approximated by FNC14 of Fig. 3.16. Figure 3.16 gives at the top the start up simulation and below the TUTSIM listing of the mechanical parts of an AC induction motor. At high slip, the electric current is very high, but this is outside a purely mechanical Bondgraph.

The model with a constant flow source and a nonlinear slip resistor is very comprehensive in the sense that it also gives the losses and efficiencies. Hence, our motor has a slip efficiency, given by the ratio of real to synchronous shaft frequency; it is practically equal to the total efficiency since the drag efficiency, due to bearing and electric losses, is near to 100%. The relation between the different efficiencies will be treated in sec. 3.5.

Finally we could mention that the other type, the AC synchronous motor is represented by a very similar Bondgraph to Fig. 3.15, only with a nonlinear spring C14 instead of R14. This allows no continuous slip but can lead to torsional oscillations between bond 12 and bond 23. The Lanchester damper for such oscillations has already been treated in sec. 3.3.

3.5 Transmissions and Efficiencies

A clutch has equal input and output torques, but a transmission is a transformer in the sense of Bondgraphs, that increases speed and reduces torque such that power is constant. Its Bondgraph appears in Fig. 3.17, on top for a fixed ratio with a TF and below for a controlled ratio with MTF (modulated transformer). The equations with the control signal α are written at the right.

Examples of transmissions in mechanical engineering are gear reducers, belt and chain drives and roller drives; the two last ones can be made adjustable. The difference of input and output torques will be going through the casing to the ground as already discussed in sec. 3.3. Let us repeat that this follows from the conservation of torque or more precisely from conservation of angular momentum.

The elements TF and MTF are power conserving and 100% efficient, since efficiency is defined as power out divided by power in when operating steadily. Any inefficiency is introduced by friction or R-elements. Here we shall consider them in parallel, leading to the slip and slip efficiency, and R-elements in series giving drag or torque loss and drag efficiency.

Figure 3.18 shows at the top a transmission with a resistor R22 in parallel, calculating slip $\omega 22$ and slip efficiency as written below. The second line of Fig. 3.18 shows a transmission with a series resistor, producing a loss torque M23 and the drag efficiency written below. The product of both efficiencies is the total efficiency. This relation holds approximately also for a transmission having both slip and drag losses, as long as these losses are not too high. These notions are important for simulation in practice, because both efficiencies are often available and help to determine the parameter values of the slip and drag resistors.

If efficiencies are not available, they can be estimated from the blocked shaft and free shaft experiment which are explained by the Bondgraph of Fig. 3.19. This

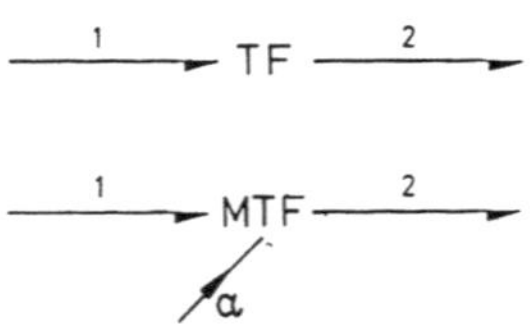

$$\omega_2 = r\,\omega_1$$
$$M_1 = r\,M_2$$
$$\omega_2 = r(\alpha)\,\omega_1$$
$$M_1 = r(\alpha)\,M_2$$

Figure 3.17 A transmission acting as a transformer in case of fixed ratio or as an MFT in case of adjustable ratio. Torque is changed according to the speed ratio r such that power is conserved. Bondgraph at the left, equations at the right.

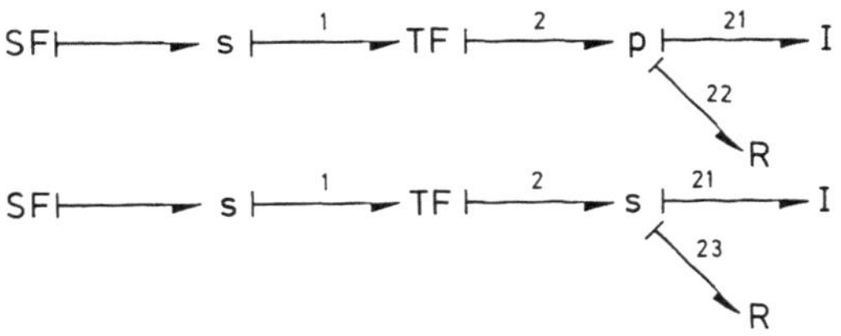

$$\eta_{slip} = \frac{\omega_{21}}{\omega_2}$$

$$\eta_{drag} = \frac{M_{21}}{M_2}$$

$$\eta_{tot} = \frac{\dot{E}_{21}}{\dot{E}_1} = \frac{\omega_{21} M_{21}}{\omega_1 \; M_1}$$

$$\eta_{tot} = \eta_{slip} \times \eta_{drag}$$

Figure 3.18 Parasitic effects added to a transmission: R22 is a slip resistor resulting in slip efficiency, R23 is a drag resistor leading to drag efficiency. The total efficiency is defined as power out divided by power in steady operation; it is the product of slip and drag efficiencies. Bondgraph at the top, equations below.

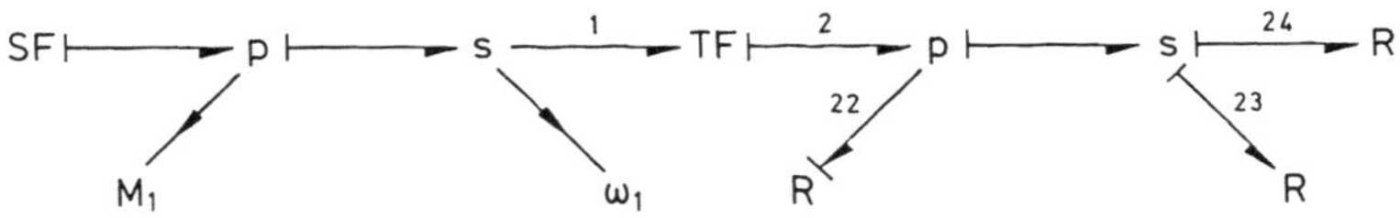

blocked shaft	free shaft
$R_{24} = \infty$	$R_{24} = 0$
$\omega_{24} = 0$	$M_{24} = 0$
$M_1/\omega_1 = r^2 R_{22}$	$M_1 \omega_1 = r^2 R_{23}$

Figure 3.19 Bondgraph for parameter determination (slip and drag resistor) from a blocked and free shaft experiment. Only measuring devices for torque and rotation frequency on the input shaft are required, and the shaft is alternatively clamped (fixed) or left alone (free).

figure refers purely to steady state without acceleration torque, hence without I-elements. Therefore it has algebraic loops and is incomputable (sec. 6.5). The signal bonds starting out from the p- and s-junctions at the left represent the measurement devices for torque and rotation frequency. The element R24 represents a variable load.

Blocked shaft: R24 is made so large that the output shaft is stopped or blocked. Any movement of the input shaft comes from the slip in R22. Starting up the drive, that is the flow source at the left (please be careful not to overload the drive), we can measure the torque M1 and speed $\omega 1$ and determine R22 from the formula on bottom left on Fig. 3.19.

Free shaft: Here R24 is set to zero, hence, M24 is zero and any input torque comes from the drag resistor R23. Starting now, the flow source R23 is determined similarly from the formula at the bottom right of Fig. 3.19.

The merit of these experiments is that they only require very simple loading provisions (R24). Free shaft means no loading device at all and a blocked shaft can be realized by tying or screwing the shaft to the ground. Of course, it is better to have more elaborate loads and instruments, but these are often not available.

There are 2 sources of error in the determination of efficiency:

1. The drive at the left must be a pure (ideal) flow source and not slow down under load.
2. In adjustable transmissions (MTF), the ratio r must not change with loading.

In water turbines and torque converters for automobiles, the ratio r changes necessarily with load. Consequently slip and drag efficiency become difficult to distinguish and much less useful than in other transmissions like gear reducers (no slip for 100% slip efficiency), belts (both slip and drag). In hydrostatic machines, they become volumetric efficiency (instead of slip) and mechanical efficiency (replacing drag). For details see sec. 4.3.

4 Simulation of Fluid Power Systems and Hydrostatic Drives

4.1 Hydrostatic Power in Fluid Flow

Since the time of Pascal (about 1650) fluid under pressure has been used for many engineering purposes, especially for hydraulic presses. Such presses are able to exercise large forces at moderate speeds, today mainly for the deformation of sheet metal in automobile bodies, earlier (1800) for compacting textiles. Since 1950 the technology is called fluid power engineering and includes the transmission and control of rotary motion in hydrostatic transmissions. Fluid power itself is defined as the technique of transmission of energy and information by fluid under pressure. Here Bondgraphs help to clarify the interaction of the many variables. For efficient design and control, the technique requires a special way of thinking, a mixture of standard mechanical engineering, hydraulics and control engineering.

Since pressures are high and flows or fluid velocities relatively low some simplifications are possible, as explained by the example of a pump in Fig. 4.1. At the top the pump is driven by the rotary bond and takes fluid in the vertical bond for delivery to the horizontal bond at the right; these bonds represent fluid intake and delivery respectively. They carry the ring indicating the flow of several power components which are:

1. The hydrostatic part represented by the static pressure p and volume flow Q.
2. The kinetic energy transport due to fluid velocity and mass flow.
3. The transport of internal energy.

There are several possibilities to select flow and effort, but the most appropriate seem to be mass flow and h, the enthalpy per mass (see chapter 7 for enthalpy). They are entered on Fig. 4.1. With incompressible fluids only the hydrostatic part, number one, is retained and parts 2 and 3 neglected; we shall return to them in section 7.2 in connection with thermodynamics of hydrostatic fluids.

Further in Fig. 4.1 bottom, the intake pressure can be taken as 0, so that the intake bond disappears because its pressure is grounded (taken as reference). The pump becomes a simple transformer with a rotary mechanical and a hydrostatic bond, that has the standard pressure and volume flow.

Speaking about the notation for volume flows we use Q as normal in fluid power engineering. International standard is $\dot{V}$, but inconvenient except for

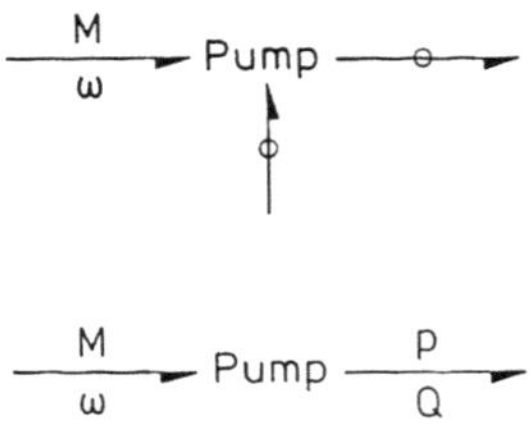

Figure 4.1 Word Bondgraph for a hydrostatic pump with intake and delivery ports, each carrying hydrostatic and other power, as indicated by the rings for multibonds. Below only hydrostatic power is retained (dynamic pressure and internal energy transport neglected), further intake pressure is set to zero.

thermodynamics (sec. 7.2). Thinking it over, we conclude that the best solution is to use both notations.

4.1.1 Hydrostatic or Hydraulic?

In principle hydraulic refers to the application of incompressible fluid mechanics and includes the effect of gravity, e.g. canals, dams and water power stations. Hydrostatic also implies that gravity forces are negligible as in fluid power technology. Hence we should only use the word hydrostatic in this chapter.

In practice fluid power engineers use both terms without a difference in meaning, more or less by habit. We follow this habit.

4.1.2 Pumps, Hydrostatic Machines and Transformers

In Fig. 4.1 we use the word pump for the device that converts mechanical into fluid power. However, in practice most are completely reversible and can indifferently work as a pump or a motor, that is transfer fluid power into mechanical power or vice versa. Therefore the name hydrostatic machine is more appropriate (abbreviated as HSM). The flow of power, pumping or motoring can instantly be reversed, often several times per second.

An ideal hydrostatic machine is then a transformer, as already stated in sec. 2.2, and represented in Fig. 4.2 at the left: On top the Bondgraph symbol, in the middle the international standard circuit symbol and at right the equations. Here the displacement, that is the volume of incompressible fluid delivered per radiant of shaft revolution, becomes the transformer modulus; our equations are those of a transformer in the sense of Bondgraphs.

In many cases, the transformer modulus or displacement can be adjusted by a control parameter. This is represented as a modulated transformer below on Fig. 4.2, further as circuit symbol and equations; α is the control parameter that varies from 0 to 1 or from -1 over 0 to 1 and D_0 the maximum displacement.

Whilst Bondgraphs and equations are simple, the real design of hydrostatic machines is complex, particularly with variable displacement. There exist gear, vane and piston machines, the last having many pistons arranged axially around

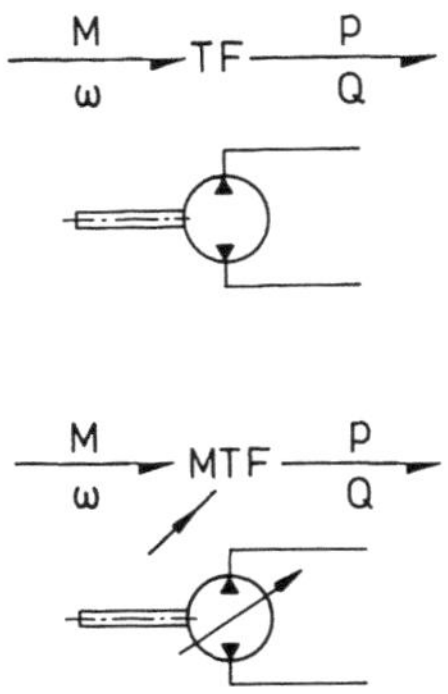

$$M = D\,p$$
$$Q = D\omega$$

$$M = \alpha D_0\,p$$
$$Q = \alpha D_0\,\omega$$

Figure 4.2 Bondgraph symbol, circuit symbol and equations of hydrostatic machines (pumps and motors). They are power conserving transformers with fixed displacement or modulated transformers with variable displacement. Any losses, efficiencies and other parasites need to be written separately.

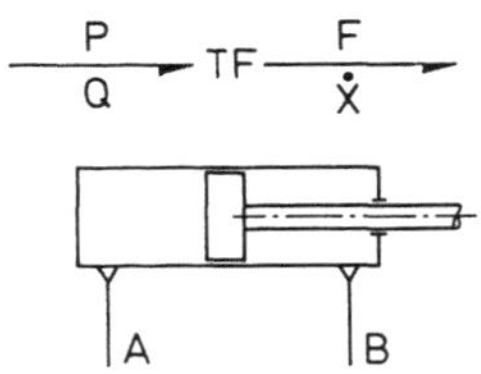

$$F = Ap$$
$$Q = A\dot{X}$$

Figure 4.3 Hydraulic ram, also a simple transformer when outlet pressure is neglected.

a shaft. Variable displacement is obtained by varying the piston stroke, inclining an internal plate or even the axis (bent axis machines). Here our control parameter α is approximately proportional to the angle of inclination (Thoma 1970 A): All these complications to realize an MTF. In addition, the realization brings in parasitic effects and inertias, losses and efficiencies as we shall see in sec. 4.3.

Figure 4.3 shows another transforming device, the hydraulic ram or cylinder which transforms hydrostatic into mechanical power, again as Bondgraph symbol, as circuit symbol and as equations. In the Bondgraph and the equations the pressure at the outflow (port B) has been neglected or set to 0, also the piston rod is assumed to have a neglegible cross-section.

4.1.3 Pressure Levels and Rotary Casings

If we allow the casing of a hydrostatic machine to rotate and admit definite pressures on both pipe connections, we observe:

1. The pumping or motoring action is proportional to the difference of rotation frequency of shaft and casing (already used in sec. 3.3).
2. The entering fluid flow equals the leaving flow and torque is proportional to pressure difference.

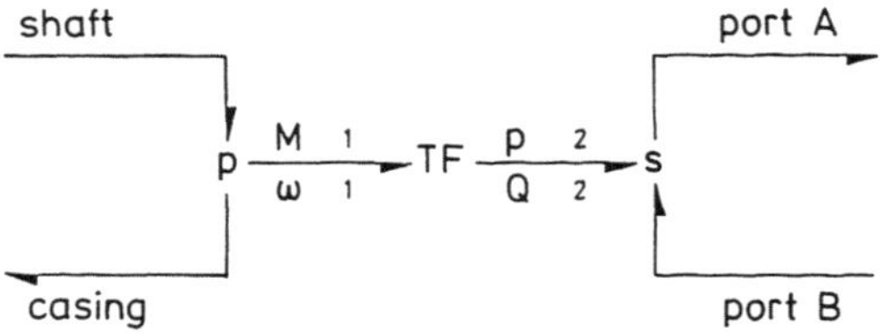

Figure 4.4 Bondgraph for a hydrostatic machine with rotary casing and pressures in ports A and B; losses are neglected. This Bondgraph is valid also for hydraulic rams with moving casing and piston rod and with pressures in both ports A and B.

Consequently the transformer TF representing the hydrostatic machine (or MTF with variable displacement) is connected by a p-junction to the bonds representing shaft and casing frequency of rotation, and by an s-junction to the bonds for intake and delivery ports. This is shown in Fig. 4.4. The equations of Fig. 4.2 then refer to the variables in bonds 1 and 2, that is net rotation frequency and net pressure. Setting the casing frequency and intake pressure to zero returns to the Bondgraph of Fig. 4.2.

The Bondgraph of Fig. 4.4 is also applicable to the ram of Fig. 4.3 with force/speed of the piston rod and casing on the mechanical side and pressure/flow of both pipe connections on the hydraulic side.

4.1.4 Hydrostatic Resistors

Hydrostatic resistors, also called throttles, appear frequently, either as desirable control elements or as undesirable parallel resistors for leakage (sec. 4.3). Physically we have two kinds of resistors:

1. Viscosity controlled resistors, where pressure is proportional to flow (linear resistor). It is also proportional to viscosity and consequently strongly temperature dependent, since viscosity decreases drastically with temperature (by about a factor of two every 20K temperature increase with normal hydraulic and engine oils).
2. Inertia controlled resistance with pressure drop proportional to flow squared, and approximately independent of temperature. This includes the so-called turbulent flow regime.

As shown in Fig. 4.5, resistors are R-elements in a Bondgraph. In fluid power there are different symbols for viscosity and inertia controlled resistors (Fig. 4.5, top left and middle), but the Bondgraph symbol is always an R-element, but we can indicate by the subscript vis and ine the character of the element.

A real piece of hydraulic equipment, be it a pipe, a valve or an undesirable leak path in a hydrostatic machine, always has both pressure drops in series, as shown on the right of Fig. 4.5. So the total pressure drop is the sum of a linear term (viscous) and of a quadratic term (inertia controlled). At very low flows the viscous term is always predominant, a fact which is important in practical simulation. The cross-over, the flow where both pressure drops are equal, depends on the viscosity and on the temperature.

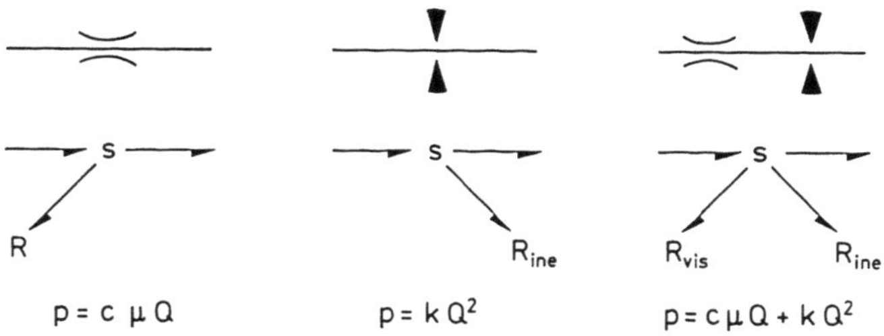

Figure 4.5 Hydraulic resistors, viscous, inertial and combined, above as circuit symbol, below as R-element. Each real pipe (or other hydraulic piece) has both effects in series (at right), so at low flows viscous pressure drop is dominant.

The values of the constant c and k in the equations of Fig. 4.5 can be taken from hydraulic manuals or from Thoma 1970 A.

4.1.5 Hydrostatic Capacitors

Capacitors and capacity effects are important in fluid power, especially in the resonance of hydrostatic servomotors. Basically any element where pressure is proportional to injected volume is a capacitor. It may be due to the compressibility of the hydraulic fluid itself, due to the expansion of the walls of a vessel under pressure, or take place in a special device called hydraulic accumulator.

Figure 4.6 shows at the top left the circuit symbol of a hydraulic accumulator, below as C-element in parallel. A special type, the hydropneumatic accumulator, has the characteristics shown in Fig. 4.6, top right: Up to a certain pressure, called filling pressure, it admits no fluid, above according to the characteristics shown, approaching the nominal volume with very high pressures. For details see Thoma

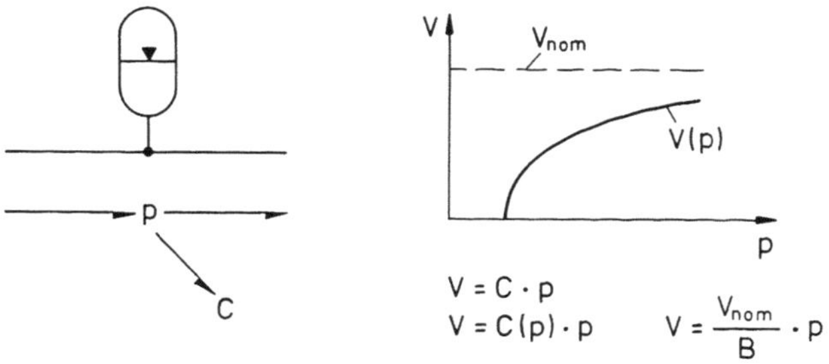

Figure 4.6 Circuit symbol, Bondgraph symbol and volume/pressure characteristic of hydropneumatic accumulators. Vnom is the nominal volume, pfill is the prefilling gas pressure.

1970 A. As shown in the equation of Fig. 4.6, absorbed volume is capacity times pressure, where C depends on pressure in non-linear cases, such as hydropneumatic accumulator just discussed. If the absorbed volume is due only to the compressibility of oil, the last equation applies, with Vnom the volume of the container or a tank put under pressure, B the compressibility modulus: It is about 1400 MPa, meaning that oil reduces its volume by 1% for an applied pressure of 14 MPa.

Figure 4.7 relates to the simulation of a hydropneumatic accumulator, that is a non-linear C-element, built from a C131 and FNC13, which models the compression of the gas when the prefilling pressure has been exceeded. Due to integral causality, input 131 is a volume and output 13 is the pressure. We have to

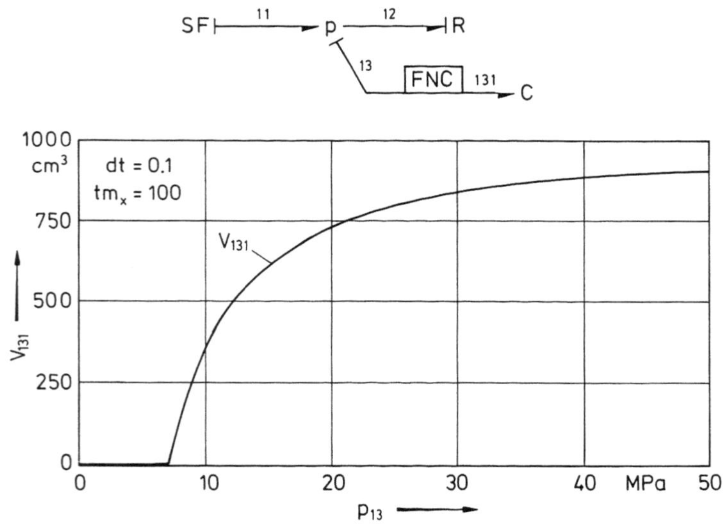


```
10.0000              11 SF       111                    ;loading flow source
 0.0330000           12 R         13                    ;bypass resistor
                     13 FNC      131
   1     0.0000               0.0000
   2     2.0000               7.0000
   3   194.0000               8.3200
   4   351.0000               9.9000
   5   478.0000              11.7000
   6   580.0000              14.0000
   7   727.0000              19.8000
   8   823.0000              28.0000
   9   926.0000              56.0000
 1.0000             111 CON                             ;file bonsi41.sim
 1.0000             131 C         11      -12           ;basic capacity
 0.0000
```

Figure 4.7 Simulation example of hydropneumatic accumulator with 7 MPa filling pressure. For numerical reasons a small initial capacity ($2\,cm^3$ at 7 MPa) is included.

admit for numerical reasons a small capacity even below prefilling pressure (7 MPa in the example), corresponding to $2\,cm^3$ absorbed. This capacity forms a time constant with other elements of the Bondgraph and determines the calculation step (sec. 6.4); with zero capacity the time step would be zero and the Bondgraph incomputable.

The graph on the bottom of Fig. 4.7 has time but not pressure on the x-axis and adsorbed volume as the y-axis; it shows the pressure/volume characteristic already indicated in Fig. 4.6. The effect of the initial capacity, $2\,cm^3$, is seen to be negligible until 7 MPa compared to the capacity in the working range, about $900\,cm^3$ for 50 MPa pressure rise.

Hydraulic accumulators of the above kind are a versatile device, used also as temporary hydrostatic power source: they come in nominal sizes from 0.2 to 400 liters. They are filled with nitrogen because of fire hazard. Oil and nitrogen gas are separated by a bladder (flexible rubber skin) or a piston. The gas filling can be done by the user to the desired prefilling value and needs to be checked about once a year because of diffusion losses. The pressure/volume curve is polytropic (between adiabatic and isothermal); we have assumed a polytropic exponent of 1.25 in Fig. 4.7. For a long bladder life they should be charged and discharged only between 25% and 75% of nominal volume.

4.1.6 Hydraulic Inertias

Inertia effects in hydraulic pipes come from the force required to accelerate a fluid with flow increase. They are explained in Fig. 4.8; on the top a schematic cross-section of a pipe, below an I-element in series and at the bottom a circuit symbol used by the author, but not standardized. The equations are shown at the right: They start with force is mass times acceleration, which leads to a relation between pressure and flow change, where A = cross-section, 1 = length of the pipe and ρ mass density (about $850\,kg/m^3$).

The value of the I-element is proportional to the length divided by the cross-section of the pipe, hence more important in narrow and long pipes. In practice the pipe inertia has to be taken into account in long narrow pipes and at high frequency of pressure and flow variations.

The above contains enough information on hydrostatic components for efficient simulation.

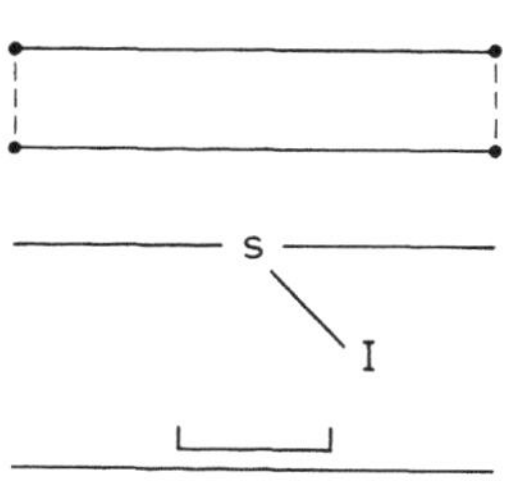

$$F = m \frac{d\dot{x}}{dt}$$

$$p = \frac{\varrho L A}{A^2} \frac{dQ}{dt}$$

$$I = \frac{\varrho L}{A}$$

$$p = I \frac{dQ}{dt}$$

Figure 4.8 Action of fluid inertia in pipes: Cross-section, I-element, in the sense of Bondgraphs and circuit element at the left, equations at the right. The inductance or intertia effect is large for long pipes with small cross-section A.

4.2 Valve Controlled Actuators and Servomotors

A basic fluid power circuit comprises a control valve and a piston in a cylinder or ram. Figure 4.9 shows on top a schematic, with a spool valve at the left connected to the P (pressure) and T (tank) ports. Depending on the position of the spool valve it passes fluid from the P to the A port and from the B to the T port; this is so with the spool upwards, whilst downwards the P port is connected to B and A to T.

At the right of Fig. 4.9 the ram appears rigidly connected to the load, represented by a circle. Depending on the spool position, the load moves to the right with the spool upwards and to the left with the spool downwards, with a speed approximately proportional to the spool displacement. In fact, the openings (windows) between the spool flanges and A and B ports are position controlled resistors, as already described in Fig. 2.29. We thus have a simple actuator; it can be turned into a servomotor by a mechanical return (not shown). Then the spool position equals the difference of an input position and the piston position.

The lower part of Fig. 4.9 is a Bondgraph including the mechanical return, where the difference of position 124 (integrated from piston speed 24) and of the input position 10 controls the flow source. The Bondgraph is simplified in that the pressure in bonds 11 to 19 represents the difference of the actual pressures of A and B ports. Further, since the mechanical return appears as signal bond (connection), any forces in it are neglected, including the substantial flow forces on the valve spool. They always act to close the valve.

The flow source SF 11 gives a valve flow proportional to pressure, R12 represents both any leakage and the reduction of the valve flow with opposing

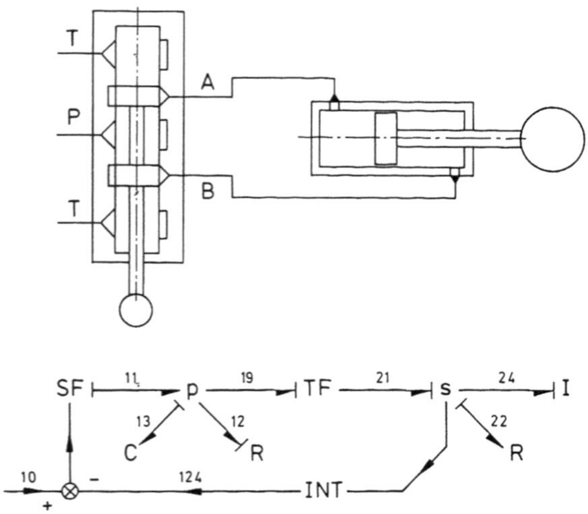

Figure 4.9 Schematic (top) of a hydrostatic servomotor with a spool valve at the left, having P (pressure) and T(tank) ports, A and B ports leading to the ram at the right. Moving the spool upwards, fluid from the P port is connected to the A port on the ram and from the B port to the T port: the piston and load move to the right. Simplified Bondgraph below.

$$Q_{11} = K \times s_{111}$$

$$Q_{12} = R_{12} \cdot p_{13} \qquad F_{21} = A \cdot p_{13}$$

$$Q_{19} = A \cdot \dot{X}_{24} \qquad F_{22} = R_{22} \cdot \dot{X}_{24}$$

$$p_{13} = \frac{1}{C} \cdot \int (Q_{11} - Q_{12} - Q_{19})\, dt \qquad \dot{X}_{24} = \frac{1}{I_{24}} \cdot \int (F_{21} - F_{22})\, dt$$

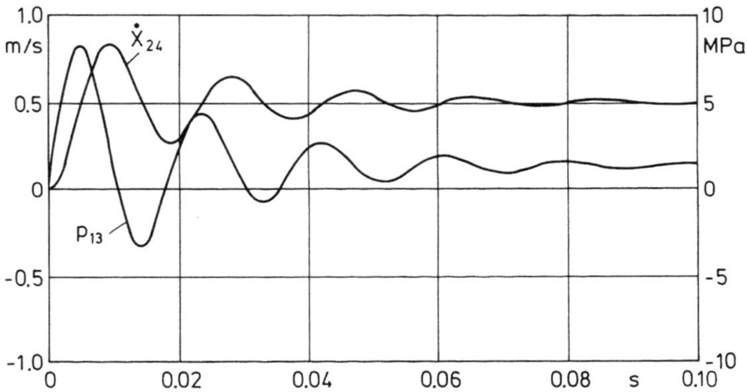


```
Timing: 500.000E-06   ,DELTA  ;    0.1000000   ,RANGE
PlotBlocks and Scales:
Format:
     BlockNo,  Plot-MINimum,   Plot-MAXimum;  Comment
Horz:      0 ,   0.0000      ,   0.1000000  ; Time
  Y1:     24 ,  -1.0000      ,   1.0000     ; load mas
  Y2:     13 , -10.0000      ,  10.0000     ; oil compressibility
  Y3:        ,               ,              ;
  Y4:        ,               ,              ;

 980.0000          11 SF     111                      ;flow source
  11.6000          12 R       13                      ;leakage
   0.3510000       13 C       11     -12     -19      ;oil compressibility
   0.0000
   1.963E+03       19 TF      24                      ;piston flow
   1.963E+03       21 TF      13                      ;piston force
   5.000E+03       22 R       24                      ;load friction
 100.0000          24 I       21     -22              ;load mas
   0.0000
   1.0000         111 CON                             ;file bonsi42.sim
```

Figure 4.10 Equations, TUTSIM listing and simulation in the time domain of the servomotor without the mechanical return. The listing is very similar to the equations above, in fact it is a kind of condensed writing of them.

pressure: It is a linearization of the equations 2.29. Capacity C13 represents the compressibility of the fluid between control valve and ram. The main contribution comes from the fluid in the volumes between the cylinder and piston in the ram, where one usually assumes the piston to the the mid position. The piston itself acts as transformer TF19,21.

On the mechanical side, the load is represented by its mass I24 and by a linear friction R22, which includes friction of the seals between cylinder and piston rod. Finally the integrator INT124 calculates piston position from piston speed, itself calculated in I24 as the causal stroke indicates.

We see that there are many simplifications, but this is the standard model used in text books (Martin 1980) for small signal operation. We see that the Bondgraph for valve and ram contains two energy storing elements (C12 and I24), hence it is a second order element damped by both resistors. The mechanical signal return contains a further integrator (INT124); consequently the overall system is of third order. Writing the equations, either directly or after transforming the Bondgraph into a blockdiagram, is simple, as is the calculation of the transfer function.

Before turning to simulation, a few words about writing the Bondgraph: The p-junction indicates that from the valve flow Q11, leakage flow Q12 and compressibility flow Q13 is substracted to obtain Q19 available for pushing the piston, whilst the corresponding pressures are equal (this is human reasoning without regard to the indicated causalities. The s-junction says that piston force is needed to overcome friction force F22 and to accelerate the load (F24), and further that all speeds are equal.

For simulation Fig. 4.10 contains the equations in standard mathematical notation for which the transfer function, say between s111 (input signal) and piston speed $\dot{x}$ 24 is easily determined by elimination of the internal variables. The TUTSIM model listing follows and it is strikingly similar to the standard equations.

Figure 4.11 shows at the top the simulation result as a function of time and below as a function of frequency, obtained by a FFT (Fast Fourier Tansform) with the FANSIM program (Ken Fyfe, Jean Thoma and F. Ismail 1986). The example is sufficient to show the principle of servomotor simulation, to which various refinements can be added with additional elements in the Bondgraph. In Fig. 6.19 we shall meet another example of servomotor simulation in connection with a derivative causality in a hydraulic robot.

In order to demonstrate how quickly a derivative causality can appear, consider the servomotor of Fig. 4.9 with the casing of the ram connected to the ground by a spring; this occurs in aircraft hydraulics when the ram is anchored on a deformable structure. Here we need the systematic writing procedure of sec. 3.3 and assign an s-junction each to the piston and casing position. Figure 4.12 contains a partial Bondgraph derived from Fig. 4.9 and has the 2 s-junctions and the transformer for the ram which acts between them through a p-junction. The

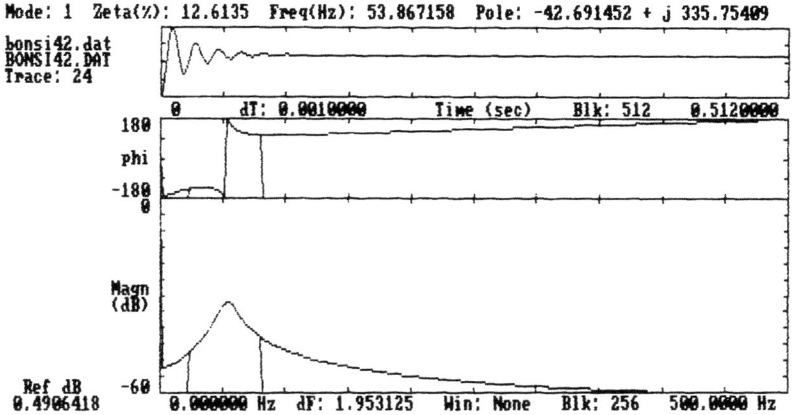

Figure 4.11 Simulation result in the frequency domain, obtained from Fig. 4.10 by an FFT with the FANSIM program.

upper s-junction connects to R22 and I24 as before. The lower s-junction could be connected through C33 to the ground to represent the deformable anchoring. This would give a causal conflict because C33 would be in parallel with C13 through the transformer.

The remedy is simple and already shown in Fig. 4.12: we add an I-element and causality goes well as shown. It represents the mass of the casing and should make no difficulty (of small parameters).

The experienced simulator would anticipate another difficulty, however. There is no friction near C33 and I34 so that these elements can represent a very weakly damped resonance. We could add an R-element (say R32) but elastic metal deformation has almost no damping: A problem in the life of a practical simulator. Perhaps he can make the R-element much larger, based on experiments, but a little of the bad conscience of sec. 6.5 remains.

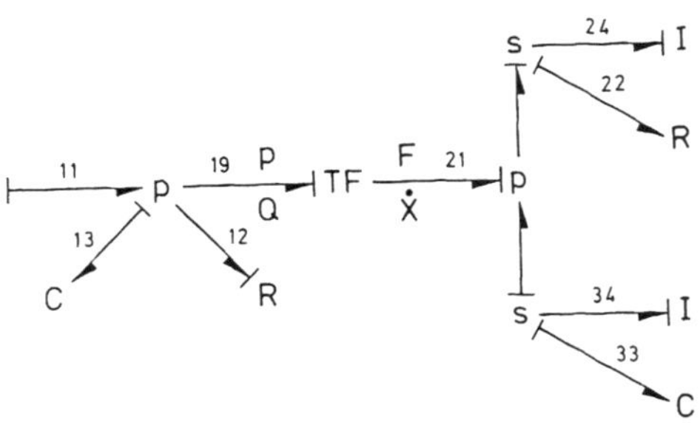

Figure 4.12 Part Bondgraph related to Fig. 4.9 with the casing of the ram connected to the ground by a spring (C33). The s-junctions represent the speeds of the piston and casing, respectively. Without I34, the mass of the casing, there would be a causal conflict because C13 and C33 are in parallel through the transformer.

4.3 Circuits and Hydrostatic Transmissions

Similarly as with mechanical devices in sec. 3.3 there is a systematic way to write hydraulic Bondgraphs according to the following procedure:

1. Assign a p-junction to all points of definite pressure.
2. Place all elements, except C-elements between appropriate p-junctions behind as s-junction.
3. Only C-elements are directly attached to the p-junction that represents the pressure at its entrance.

Here C-elements always work between a hydrostatic pressure and ambient pressure; the very few exceptions are outside the scope of this book. The resulting Bondgraph can be simplified by grounding one p-junction which means that all other pressures are counted from there.

4.3.1 Hydrostatic Transmission Example

In its usual version, a hydrostatic transmission has a variable displacement machine, driven by an engine or other prime mover and is shown in Fig. 4.13, top, as a circuit with standard fluid power symbols. It delivers fluid to a similar fixed displacement machine that drives the output shaft. The fluid returns through the

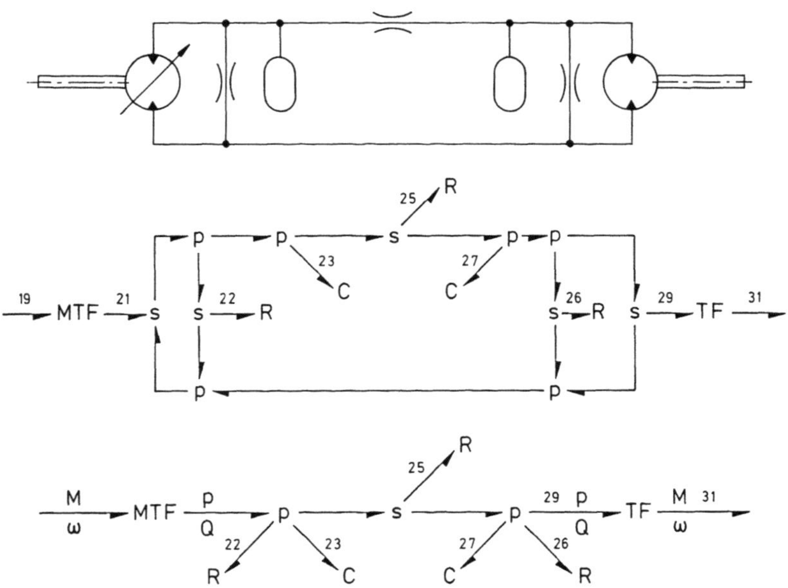

Figure 4.13 Hydrostatic transmission comprising a variable primary and fixed secondary hydrostatic machine, resistors R22 and R26 for leakage, R25 for resistance of the main hydraulic conduit and C23, C27 for compressibilities. Circuit on top, systematic Bondgraph in the center and simplified Bondgraph below, obtained by grounding the lower p-junctions.

lower line to the variable primary hydrostatic machine. Some necessary auxiliary valves are not shown in Fig. 4.13, but several capacitors and resistors for parasitic effects.

Figure 4.13 contains in the center the corresponding Bondgraph written according to the systematic procedure, with some important bonds already labelled. A great simplification results in Fig. 4.13, bottom, if the return line is grounded with all its p-junctions. They disappear, since they have no power, whence most s-junctions have only 2 bonds and disappear until the simple Bondgraph of Fig. 4.13, bottom, emerges. After causal assignment it will be ready for computation.

4.3.2 Hydrostatic Machines in Detail

The essential action of hydrostatic machines is a transformer between mechanical and hydrostatic power, or an MTF (modulated transformer) in case of adjustable or variable displacement, as already used in Fig. 4.13. Let us note again that an MTF, and with it the essential actions in hydrostatic machines, are completely reversible. More precisely the MTF has both signal and power reversibility, described in sec. 5.3 with electric machines.

Reversibility is disturbed or removed by the parasitic effects as shown in Fig. 4.14 for a hydrostatic machine with the corresponding efficiencies written below.

The parasites of Fig. 4.14 act as follows:

- R12 is the shaft friction responsible for drag efficiency (sec. 3.5), called mechanical efficiency in hydraulic engineering.
- I14 is the moment of inertia of the shaft.
- R22 is the parallel resistance for leakage: A part of the ideal flow escapes through the clearances as leakage.
- C23 is the compressibility of the oil in the outlet port and nearby fluid conduit.
- The small MTF at the top represents the fact that many hydrostatic machines have a displacement volume slightly dependant on shaft position, calculated by INT141.

The last parasitic effect is usually described as delivery flow fluctuation with shaft position; it is easy to forget that the effort equation of the small MTF also gives a torque fluctuation with constant pressure, which can excite torsional vibrations. In the case of piston machines, one always uses many pistons to reduce displacement variations which are $+/-5\%$ in a machine with 7 pistons (Thoma 1970 A).

The total efficiency is the product of slip and drag efficiencies and depends on the values of R12 and R22. In practice these efficiencies might be known and the formulae in Fig. 4.14 serve to determine the unknown values of the R-elements from the efficiencies. Of course, they can be non-linear and it is a modelling decision on how many effects to include.

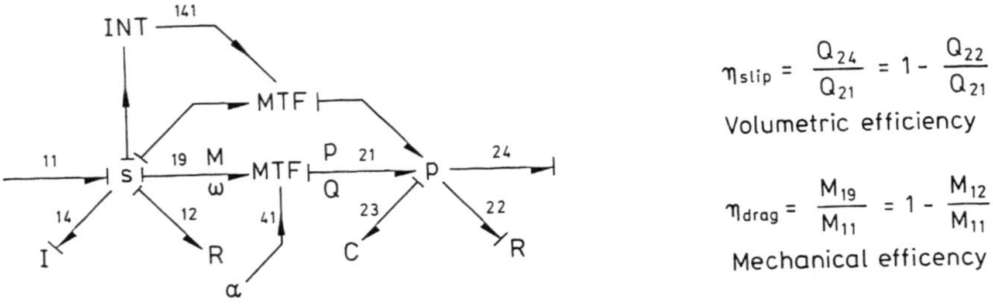

Figure 4.14 Bondgraph for a hydrostatic machine with many parasitic effects, together with formulae for volumetric and mechanical efficiencies, which are a consequence of the R-elements R12 and R22, respectively. The second MTF on top models the displacement variation with shaft position s114; it gives a cyclic flow variation at constant rotation frequency and cyclic torque variation at constant pressure.

Figure 4.15 shows an example of a variable displacement machine designed by Jean Thoma. It is an axial piston machine with 7 cyclinders arranged axially around the central shaft. Variation of the displacement is obtained by varying the piston stroke, itself by changing the inclination of the cross plate against which the pistons push (swash plate = chaban gata in the professional Japanese terminology). For details see Thoma 1970 A.

Returning to Fig. 4.13 the slip efficiencies have been included, also oil compressibility by C22 and C27, whilst mechanical parasites are excluded. If we were to neglect R25, which represents the hydraulic resistance in the fluid conduit, both p-junctions could be combined, the capacitors and leakage resistors added and we would arrive at a Bondgraph almost equal to Fig. 4.9 of the valve controlled servomotor. In fact, MTF19, 21 becomes a controlled flow source if the torque relation is ignored. This can be done if the hydrostatic machine is driven by a perfectly regulated prime mover, itself a mechanical flow source.

As a simulation example, Fig. 4.16 contains at the top a Bondgraph for a hydrostatic transmission, where I24 for the hydraulic inductance (sec. 4.1) has

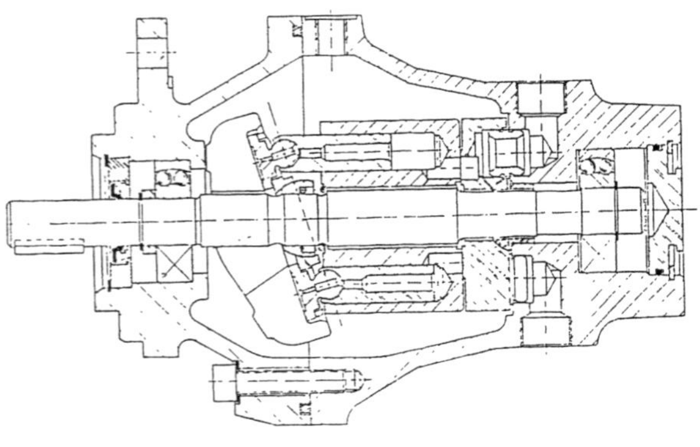

Figure 4.15 Cross-section of an axial piston machine with 7 pistons and variable displacement by the inclinable swashplate. The pistons push against the swashplate through slippers in an arrangement similar to Fig. 4.22.

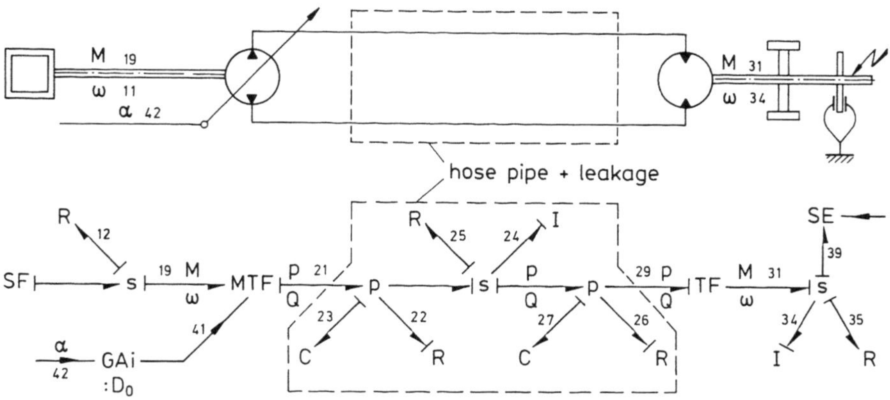

Figure 4.16 Bondgraph and circuit of a hydrostatic transmission similar to Fig. 1, with pipe inductance I24 added, also other parasites, R12 for primary shaft friction, R35, I34 for secondary friction and inertia, SE39 for constant load torque, f.i. a vehicle climbing a hill.

been added, also a load I34, R35 with SE39 modelling a force independent of speed, such as a road slope for a vehicle. The elements C23 and C27 each contain the compressibility of half of the fluid conduit or pipe, which was a rubber hose in this case. The hose is really a distributed parameter system, which would require a partial differential equation for simulation (sec. 8.2); here it is modelled by the elements in the broken lines.

Figure 4.17 finally contains the TUTSIM listing and a simulation of start up. The parameters correspond to a real application with a hydrostatic machine of $16\,cm^3/rad$ displacement. Since the Bondgraph has 4 energy storing elements (C23, I24, C27 and I34) we expect 4 Eigenvalues or 2 separate resonance frequencies, which are well visible on the simulation. We shall estimate the corresponding Eigenvalues in sec. 6.4.

4.4 Special Components and Devices

Some widely used fluid power devices can be well modelled by Bondgraphs; we include here relief valves, flow control valves and hydrostatic bearings.

4.4.1 Relief and Flow Control Valves

Relief valves have been known for over 200 years in steam boilers as safety valves: with excessive pressure they open a passage to release steam into the environment for safety. Essentially they are non-linear resistors or R-elements in parallel with a pipe, as shown in Fig. 4.18, top left. The resistor is closed at low pressure, except for undesirable leakage, up to a certain limiting pressure, when it opens and rapidly increases flow. See curve RLV in Fig. 4.18 below.

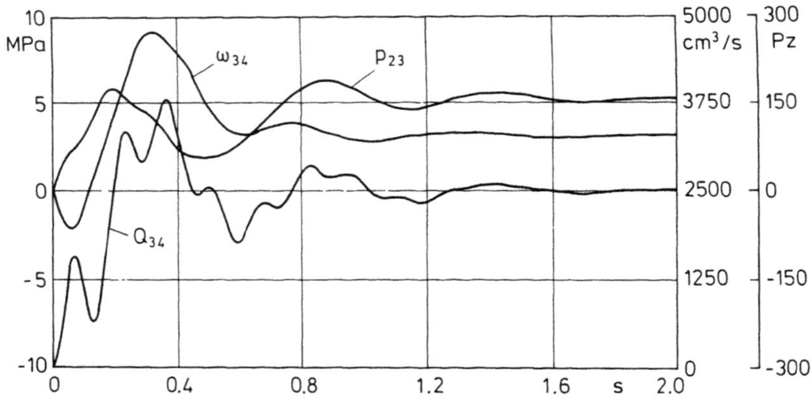


```
Timing:     0.0020000   ,DELTA   ;    2.0000        ,RANGE
PlotBlocks and Scales:
Format:
     BlockNo,   Plot-MINimum,   Plot-MAXimum;   Comment
Horz:      0 ,     0.0000     ,    2.0000     ; Time
  Y1:     23 ,  -10.0000      ,  10.0000      ; pipe compressibility
  Y2:     24 ,     0.0000     ,    5.000E+03  ; pipe inductance
  Y3:     34 ,-300.0000       , 300.0000      ; load inertia
  Y4:        ,                ,               ;

 157.0000           11 SF      111                    ;drive rota frequen
   0.0400000        12 R        11                    ;primary friction
                    19 MTF      41     23             ;primary torque
                    21 MTF      41     11             ;primary flow
   2.5000           22 R        23                    ;primary leakage
  47.0000           23 C        21    -22      -24    ;pipe compressibility
   0.0000
  26.600E-06        24 I        23    -25      -27    ;pipe inductance
   0.0000
 147.000E-06        25 R        24                    ;pipe resistance
   2.5000           26 R        27                    ;secondary leakage
  47.0000           27 C        24    -26      -29    ;pipe compressibility
   0.0000
  16.0000           29 TF       34                    ;secondary flow
  16.0000           31 TF       27                    ;secondary torque
   0.0180000        34 I        31    -35      -39    ;load inertia
   0.0000
   0.0800000        35 R        34                    ;load friction
  32.0000           39 SE      391                    ;mountain (slope) torque
  16.0000           41 GAI      42                    ;maxi displacement
   1.0000           42 CON                            ;displacement setting
   1.0000          111 CON                            ;file bonsi43.sim
   1.0000          391 CON
```

Figure 4.17 Simulation and TUTSIM listing corresponding to Fig. 4.16; since the Bondgraph has 4 energy storing elements, we expect 4 Eigenvalues or 2 resonances (sec. 6.2) which are well visible.

Near the origin, relief valves should pass no flow and flow control valves much flow as shown by the curve. A check valve is a special kind of relief valve where the pressure limit is set to zero. It passes flow in the forward direction and blocks it backwards.

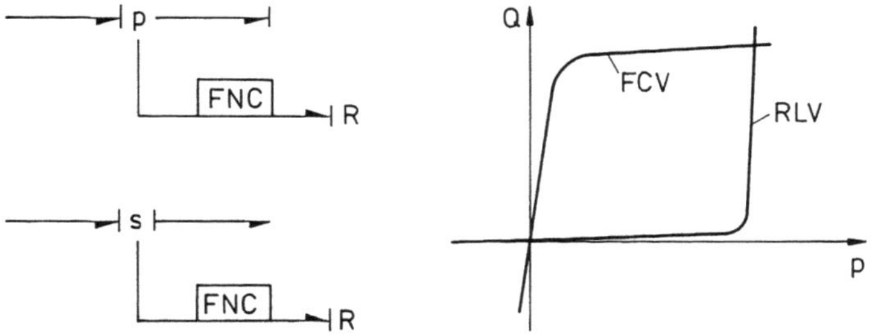

Figure 4.18 Relief (or safety) valve as nonlinear resistor in parallel to a pipe (top left). A flow control valve is a non-linear resistor in series (top right). Both are represented as an R-element with an FNC block on the flow side for the nonlinearity. Below the flow-pressure characteristics of these valves. In this simple form, the flow follows pressure instantly, that is without delay or dynamics.

Flow control valves are the counter part of relief valves, with a rapidly increasing flow at small pressures up to a flow limit, whence a large pressure rise produces only small flow increase: another typical nonlinear resistor (curve FCV). Normally such valves are put in series in a fluid conduit (Fig. 4.18, left). On the graph of Fig. 4.18, bottom, a viscosity controlled resistor (leakage) would appear as a straight line and an inertia controlled resistor as a typical root curve (sec. 4.1).

Figure 4.19 shows at the top a cross-section of a simple relief valve with a vertical piston: When the pressure exceeds the value corresponding to the spring

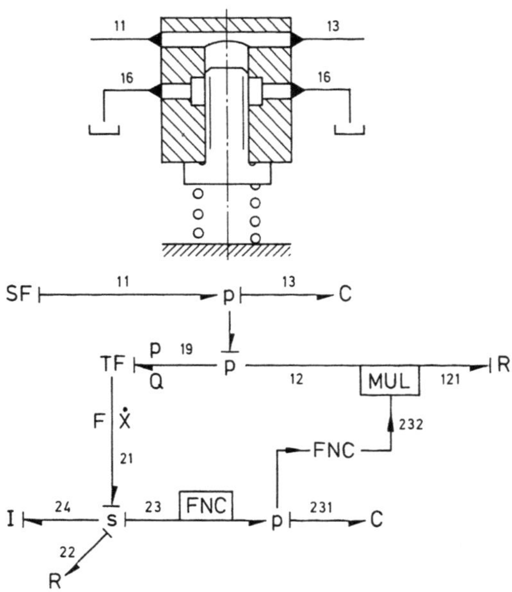

Figure 4.19 Cross-section of a direct relief valve with piston forced downwards to open a passage between conduit 11 to 13 and port 16 when pressure limit is reached. Below a Bondgraph with both mechanical and hydraulic parts, where R12 is modulated by the piston position through the MUL block. This Bondgraph includes valve dynamics and can become unstable if insufficiently damped by R22, the mechanical friction.

force, the piston slides downwards and opens a passage to pipe 16. In the case of a relief valve, the spring would be made very light so that fluid can pass from pipe 11 to pipe 16 with almost no resistance, but from pipe 16 to pipe 11 it will be blocked by the piston going upwards.

The lower part of Fig. 4.19 contains a simplified Bondgraph, where pressure in pipe 16 is set to zero. Essential for the operation are the 2 following effects:

1. The face of the piston acts as a transformer TF19,21 and produces a force on I24 and R22, representing mass and friction of the piston.
2. The piston acts as a displacement modulated resistor when it opens a passage to pipe 16. This is reticulated here as R121 multiplied by x231 or x232.

In more detail, the spring with stops is reticulated by C23, that is by C231 with a parameter of 1 so its output is numerically equivalent to position x231, followed by a function generator FNC23 representing the non-linear position/force characteristic.

Function 232 is special in that its output x232 equals x231 for positive input and is zero for negative input. In fact, the spring and stop in C23 cannot be made infinitely rigid for numerical reasons (sec. 6.2). So a negative pressure would give a positive flow by the MUL12 operator, this flow would go against pressure, a physically impossible result in an R-element (Fig. 2.17). In fact, flow will be zero neglecting leakage, as assured by FNC232. Ignoring the correction is a serious modelling error and difficult to detect since there is nothing wrong with the computer code; it can lead to violent oscillations.

Looking on the Bondgraph of Fig. 4.19, we see that the jet forces on the piston have been neglected. They act on the lower s-junction and always tend to close the flow passage and would be easy to incorporate. In practice they tend to damp the valve.

Flow control valves have a very similar Bondgraph, only that the upper p-junction in Fig. 4.19 would become an s-junction, as we might expect from Fig. 4.18.

In practice, relief valves tend to become unstable if the spring is too soft, made in order to obtain a very steep characteristic on Fig. 4.18 after opening. To avoid this and a heavy spring for large valves, piloted relief valves are available. They are a combination of the simple, direct acting relief valve above and a parallel spool valve similar to Fig. 4.9. A Bondgraph would be easy to write.

Figure 4.20 is an example of simulation of a direct acting relief valve from Fig. 4.19 with the TUTSIM listing below. Pressure increases linearly as C13 (representing the load) is filled by the flow source SF11. After 25 msec, the valve opens and flow Q12 begins. It has a large overshoot and is weakly damped. Better damping results from increased mechanical friction R12 or from making spring C23 stiffer. Unfortunately, such modifications make for less flow/pressure gain or a less steep characteristic in Fig. 4.18. Box 4.1 contains the equations corresponding to the TUTSIM listing for comparison.

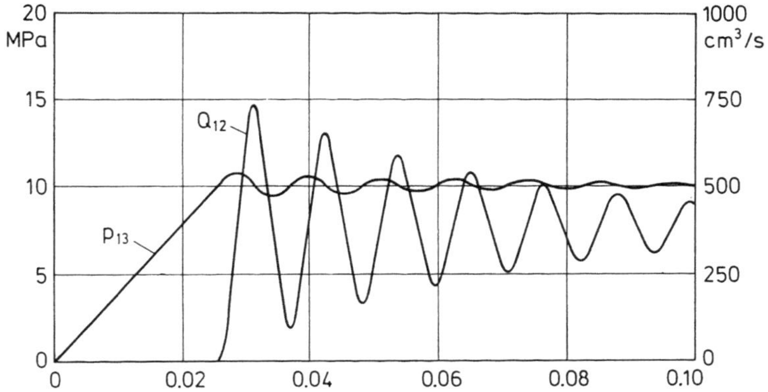


```
Timing:  50.000E-06  ,DELTA  ;   0.1000000  ,RANGE
PlotBlocks and Scales:
Format:
     BlockNo,  Plot-MINimum,  Plot-MAXimum;  Comment
Horz:     0 ,   0.0000     ,   0.1000000  ; Time
  Y1:    13 ,   0.0000     ,  20.0000     ; capacitive load
  Y2:    12 ,   0.0000     ,   1.000E+03  ; flow modulation
  Y3:       ,              ,              ;
  Y4:       ,              ,              ;

 400.0000          11 SF      111                   ;inflow
                   12 MUL     121     232           ;flow modulation
   1.0000          13 C        11     -12     -13   ;capacitive load
   0.0000
   0.5020000       19 TF       24                   ;piston (face) flow
   0.5010000       21 TF       13                   ;piston force
   0.0100000       22 R        24                   ;piston friction
                   23 FNC     231
     1   -0.4000000    -75.0000
     2    0.0000         5.0000
     3    0.4000000      5.4000
     4    0.8000000     85.4000
   3.000E-06       24 I        21     -22     -23   ;piston mass
   0.0000
   1.0000         111 CON
 600.0000         121 R        13                   ;basic flow admittance
   1.0000         231 C        24                   ;basic spring
   0.0000
                  232 FNC     231
     1   -1.0000        0.0000
     2    0.0000        0.0000
     3    1.0000        1.0000
```

Figure 4.20 Simulation of the direct relief valve of Fig. 4.19. It opens when SF11 has filled C13 to the limit pressure. The system is stable but weakly damped.

4.4.2 Note on Hydraulic Control Resistors

Hydraulic control resistors are essentially devices with variable orifices that appear frequently as above and in spool valves with four orifices controlled by the spool position. Apart from Fig. 4.19 we find them in the water pump of Fig. 9.1.

Box 4.1 Equations corresponding to Figs. 2 and 3. Equations of the direct acting relief valve corresponding the Bondgraph of Fig. 4.10 and the TUTSIM listing of Fig. 4.20.

The equations are contained in the TUTSIM listing of Fig. 3 but we rewrite them in standard mathematical notation for comparison.

$$Q_{11} = \text{const} \qquad Q_{12} = Q_{121} * x_{232}$$

$$p_{13} = (1/C) * \int dt(Q_{11} - Q_{12} - Q_{13}) \qquad Q_{19} = A * x_{24}$$

$$F_{21} = A * p_{13} \qquad F_{22} = R * \dot{x}_{24}$$

$$F(23) = F(x_{231}) \qquad \dot{x}_{24} = (1/m) * \int dt[F_{21} - F_{22} - F_{23}]$$

$$Q_{121} = (1/R) * p_{13} \qquad x_{231} = \int dt\, \dot{x}_{24}$$

$$x_{232} = F(x_{231})$$

Physically they obey the orifice equation with flow proportional to window (opening) area and the root of the pressure drop.

Before we had completely linearized the spool valve (Fig. 4.10), but here only the flow/pressure relation of orifices is taken linear, and the valve remains nonlinear by the multiplication with the spool position. Why did we not use the root characteristic with the TUTSIM block GSQ?

Root characteristics often lead to trouble, namely uncontrolled oscillations at very low flows: they only appear in the simulation not in reality. The numeric reason is clear: a root characteristic has infinite conductance at the origin, that is very small resistance. So it would require a very small time step according to sec. 6.2. In fact, on Fig. 4.19 there is a loop from C23 via bonds 21 and 19 to R12, hence a time constant proportional to the resistance.

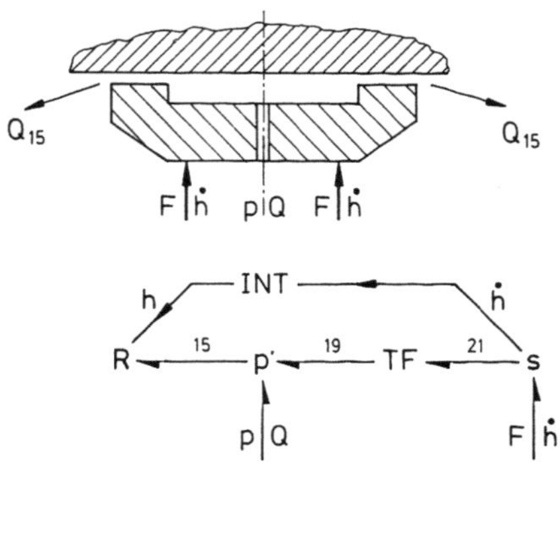

Figure 4.21 Cross-section of a round slipper acting against an upper support as prototype of a hydrostatic bearing. Oil is ducted through the central bore into the central cavity and escapes through the gap between the lips (around the central cavity) and the support, as represented by the displacement modulated resistor R15.

Such trouble does not always happen, but it is difficult to diagnose. A remedy would be to introduce a finite conductance at the origin; this could be the FNC block approximating the root characteristic, but in most cases the above pressure linearization is satisfactory.

Thinking further about the physics involved, any real hydraulic resistor has a finite conductance near the origin, that is a linear pressure/flow relation. So using a root characteristic is faulty modelling and the computer uses it to make trouble. The situation illustrates how finely physics, here fluid mechanics, is interrelated to practical computation. Let us note finally that the ABS (absolute) block has a similar, but weaker tendency with its sharp corner at the origin.

4.4.3 Hydrostatic Bearings

Another important hydromechanical device is the hydrostatic bearing, exemplified by the simple slipper on a plane in Fig. 4.21. The slipper is around the indicated vertical axis and presses under an upward force against the fixed support above; it has a central cavity surrounded by a ring or lips that can contact the support plane. To avoid such contact, oil is ducted into the central cavity from below. A pressure cushion builds up, the slipper separates and oil escapes radially from the cavity across the lips. This is called a hydrostatic bearing where the lips float in a small distance (10 to 50 micrometers) from the support, similarly as a hover-craft above the surface of the sea.

Analyzing the operation more precisely, the cavity with lips operates according to the following 2 effects, similarly as described with the relief valve above:

1. The face of the slipper works as a TF19,21. The effective area approximately corresponds to the mean diameter of the lips.
2. The space between lips and support is a displacement modulated resistor (R15 in Fig. 4.21).

The fairly complex interaction of the fluid film between lips and support is thus reticulated by the simple Bondgraph of Fig. 4.21 below. The conductance (reciprocal resistance of R15) is proportional to h^3, the third power of the gap height or separation between lips and support according to the well-known gap formula (Thoma 1970A); it is viscosity controlled and strongly depends on temperature.

The purpose of the mechanism of Fig. 4.21 is to have the slipper floating in a small distance from the support in spite of load variations, in order to avoid metallic contact with friction and wear on lips and support. The main idea is to supply an oil flow through a resistor, to be called R12, into the vertical pipe or bond of Fig. 4.21. Consequently if the pressure cushion in the central cavity is too large, the slipper separates, conductance of R15 increases such that the pressure tends to decrease again due to the increased pressure drop across R12. It is thus a kind of automatic regulation of gap height against load variations. In some cases, especially in really large bearings, R12 can be replaced by a flow source, that is a separate pump.

Figure 4.22 shows a little more complex mechanism with an upper support with mass I24. The cavity of the slipper has a capacity C13. The slipper is loaded below by the force from a big piston with a spherical contact surface allowing misalignment. Oil is supplied from a pressure source SE11 which also presses the big piston upwards through TF18,31. Further, it forces the flow through the central bore into the cavity of the slipper. The upstream resistor R12, essential for the selfregulating action, is built from a small cyclindrical part in a bore of the big piston. Its conductance is realized by the diametral clearance which can be precisely manufactured. Naturally other realizations of the upstream resistor are possible, like small orifices, but they are difficult to manufacture and to calibrate with sufficiently high resistance; further they tend to become blocked by impurities in the oil.

The Bondgraph of Fig. 4.23 is written systematically by assigning an s-junction with I24 for the speed of the upper support and an s-junction with I34

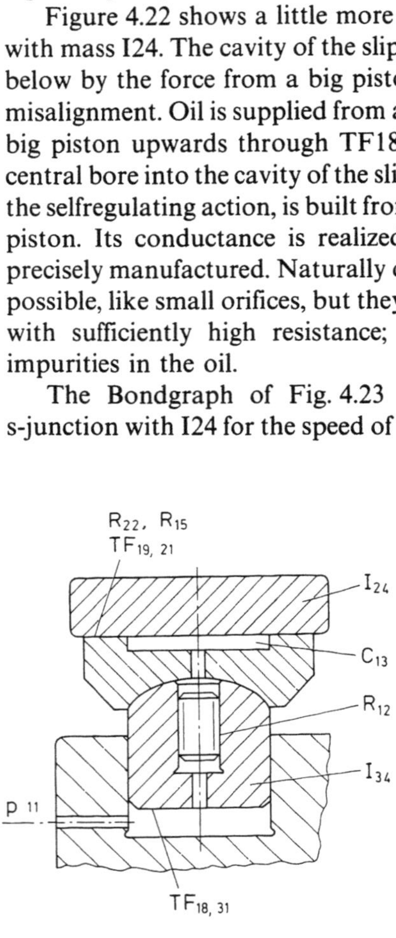

Figure 4.22 Hydrostatic bearing mechanism similar to Fig. 4.21, where the upward load force comes from the piston face, (denoted TF18,31). The upstream resistor R12 is realized by a cylindrical part fitted into a bore in the piston. The spherical contact surface allows misalignment between piston and support, which must be strictly parallel to the slipper face (self-alignment).

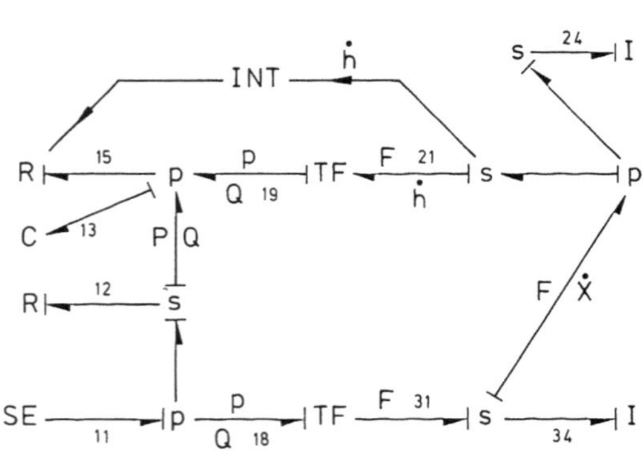

Figure 4.23 Bondgraph corresponding to Fig. 4.22 with intertia I24 for the upper support and I34 for piston (and slipper). The lifting action of the oil film between slipper and support is coupled in by the p-junction at the right. Oil compressibility C13 in the simple cavity has been added.

for the speed of the slipper with the large piston. The mechanical side of TF 19,21 is coupled in by a p-junction, since it tends to separate slipper and upper support with equal force. The hydraulic side contains fluid supply SE11, upstream resistor R12, modulated resistor R15 and the compressibility of the oil in the slipper cavity by C13, already necessary for causality. Variations of this mechanism are possible:

1. The upper support can be fixed in space (I24 removed or given infinite mass) whence we have the piston slipper arrangement of the hydrostatic machine of Fig. 4.15.
2. The upper support can be a flange on a big shaft, f.i. for water turbines or ship propellers, where many pistons and slippers carry large loads.

The Bondgraph in Fig. 4.21 serves not only to calculate variations of gap height under variable or vibrating loads, but also to illustrate the intricate working of hydrostatic bearings and hydrostatic machines in general. We are here touching the teaching value of Bondgraphs, as indicated in Fig. 3.1, as new ideas, devices or systems.

5 Electric Circuits, Drives and Components

5.1 Electric Circuits and Components

We shall treat here circuits and components and further in sec. 5.2 electric motors from an elementary standpoint, sufficient for most applications. In sec. 5.3 we shall examine the interaction of electric and magnetic fields with mechanical members in more detail. In this section we shall neglect parasitic effects of resistors, capacitors and inductors and treat them as R-, C- and I-elements. Thus, we neglect effects like inductivity of a resistance, leakage of a capacitor and winding capacity of an inductor. Such effects can always be added from the known electronic equivalent circuits once the systematic writing of Bondgraphs is familiar.

Systematic writing of electric Bondgraphs is similar to hydraulic Bondgraphs, but simpler since the question of kinetic and internal energy of fluid flow (Sec. 4.1) does not arise. Further there are no I-elements attached to inertial space as in mechanics (sec. 3.3) and no C-elements attached to absolute pressure in hydraulics (sec. 4.3). All electric elements are applied between p-junctions behind an s-junction. So the systematic electric Bondgraph writing is really simple:

1. Assign a p-junction for each point of definite voltage.
2. Place all elements between appropriate p-junctions behind an s-junction.

The resulting Bondgraph should be simplified by grounding one p-junction or a set of directly connected p-junctions.

Figure 5.1 refers to a simple electric circuit. Each wire has a definite voltage and is represented by a p-junction. The elements (R, I, C) are each placed between these p-junctions behind an s-junction, whence the Bondgraph in the middle appears.

For simplification, the lower p-junctions of Fig. 5.1 are grounded and consequently some junctions with only 2 bonds disappear. Then R12 and I14 can be combined and the much simpler Bondgraph on Fig. 5.1, bottom, emerges. Here elements are labelled from SE11 to R16 and these labels also applied to the systematic Bondgraph and the circuit. This procedure, called retro-labelling frequently helps with understanding.

It is worth noting that the systematic Bondgraph of Fig. 5.1 still is similar (topologically equivalent) to the circuit. Such similarity is lost by the grounding of

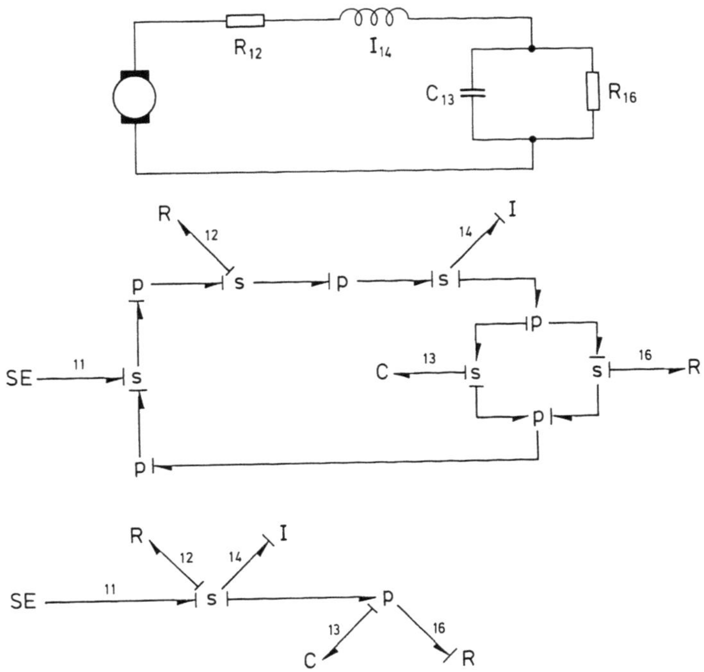

Figure 5.1 Simple electric circuit at the top. Center: Systematic Bondgraph written by giving a p-junction to each wire and placing the elements behind an s-junction. Botton: Simplification by grounding the lower p-junctions and eliminating redundant junctions.

the p-junctions in exchange for the great simplification. The relation between topology and simplicity is worth further study.

Figure 5.2 shows a slightly more complex electric circuit with a transformer. From the circuit at the top the Bondgraph in the middle is easily written by the systematic procedure. The labelling gives the numbers 11 to 19 to the bonds on the left of the transformer and 21 to 29 to those on the right; useful in more complex examples. Of course not all available labelling numbers have been used. This will allow the addition of more elements later, if needed, without disturbing the labelling scheme.

The simplification now has the special point that both sides of the transformer have a set of p-junctions that can be grounded separately and independently. This corresponds to the fact in electrical engineering that the sides of an electric transformer can have different DC potential (voltage) with respect to ground. As an example, we could have grounded the upper junctions on the right of the systematic Bondgraph on Fig. 5.2. Here it would have led to the same simplified Bondgraph, but in general this is not so with more complex circuits. Study of these questions leads deeply to the underlying mathematical structure of Bondgraphs.

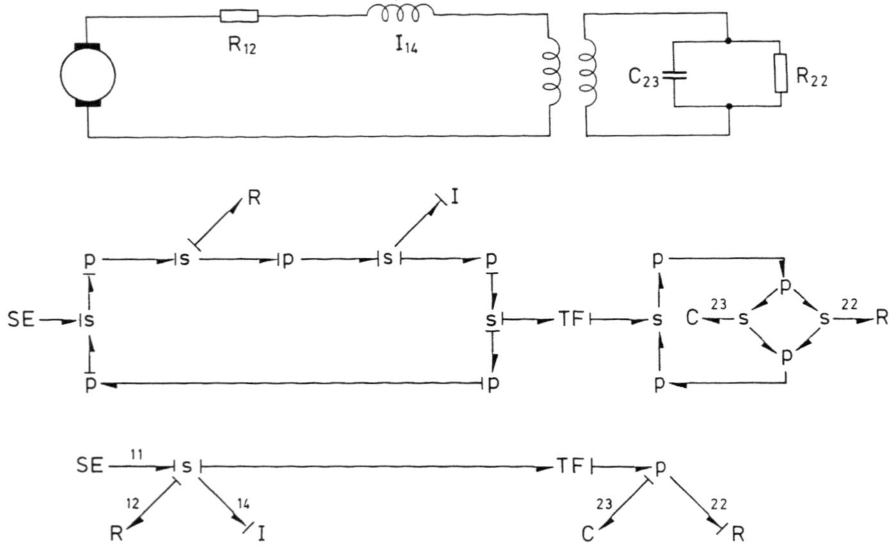

Figure 5.2 Slightly more complex circuit. Here the p-junctions on both sides of the transformer can be independently grounded. Circuit at the top, systematic Bondgraph in the center and simplified Bondgraph below.

5.2 Electric Motors

We are speaking here mainly about electric DC (direct current) motors which can be beautifully described by simple electro-mechanical Bondgraphs. Such Bondgraphs will not only allow to calculate the dynamic interaction of mechanical and electric variables, but also to explain the different motor types to the non-specialist. We elaborate here on Figs. 2.10 and 3.7, where an electric motor was represented by a gyrator. Alternating current motors of the induction type have already been treated in sec. 3.4 and we will here add a short treatment of the synchronous electric motor for alternating current.

Electric motors are based on the induction law, where a wire with electric current placed into a magnetic field (perpendicularly to the wire) experiences a sideward force; further a voltage is generated if the wire moves. These relations are written in Fig. 5.3, botton left, and are those of a gyrator, with current and speed as flows and voltage and force as efforts. Here B is the magnetic field strength that can often be varied.

The induction law is completely reversible and so are electric motors. This means that they can transform electrical power into mechanical power, as a motor, or mechanical power into electrical power as a generator. More precisely they should be called electric machines.

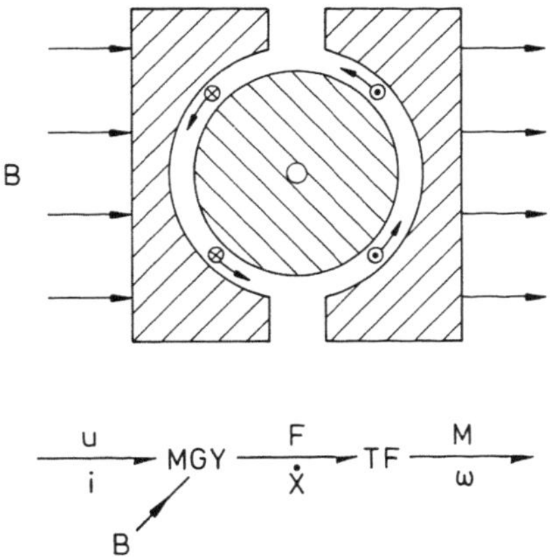

$F = BLi$
$u = BL\omega$
$M = rF$
$\dot{x} = r\omega$

Figure 5.3 Schematic of a DC electric machine with magnetic field from left to right. The windings rotate in the air gap between stator and rotor and experience a sideward force as written below. The Bondgraph in the center models this as MGY, followed by a TF that transforms force and speed into torque and shaft frequency, just like a wheel. B = magnetic field, r = radius.

Let us stop for a moment to think about the physical aspects of reversibility. In fact, in electric machines, as in hydrostatic machines, we have two kinds of reversibility:

1. The reversibility of rotation frequency and current, or in general the fact that all efforts and flows can take positive or negative values. It can be called signal reversibility.
2. The reversibility of power, which can go from the electrical to the mechanical side (motor action) or vise versa (generator action). This power reversibility corresponds to the reversibility in the sense of thermodynamics.

Further thinking and research about reversibility is a field for the future.

Returning to engineering, Fig. 5.3 is a schematic of an electric machine with a stator, the large rectangle with a big bore, where a round member, the rotor or armature is placed. It carries a winding of many wires of which 4 are shown. The magnetic B-field arrives from the left and leaves to the right; it is reconducted (to make a closed magnetic circuit) and generated by electric means not shown. Technological details, like the commutator needed to put the electric current into the moving wires, are not shown.

The center of Fig. 5.3 contains a Bondgraph with the induction effect represented by a modulated gyrator, followed by a transformer that acts essentially like a wheel (radius r), changing the circumferential motion of the electric wire into rotation. The lower part contains the equations. Since the magnetic field can be varied in general we have here a modulated gyrator. The transformer for the wheel action can be combined with the modulated gyrator, as we shall do from Fig. 5.4.

The magnetic field can be generated by a permanent magnet whence it is constant in practice, or by an electromagnet, also called field coil. Here the magnetic field is approximately proportional to the field current, but it depends on how this current is determined. It is constant, if the field current comes from a

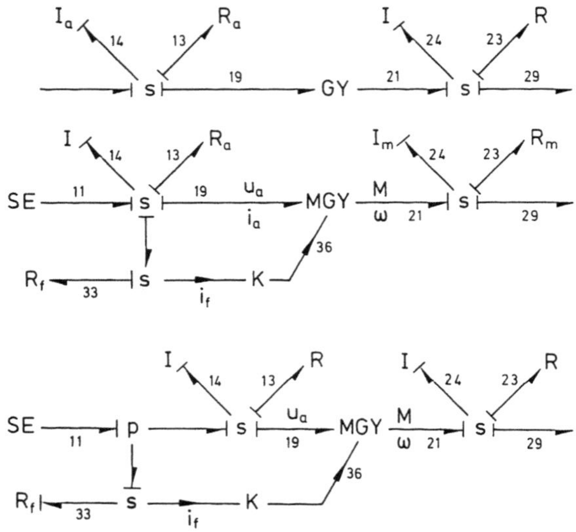

Figure 5.4 Bondgraph for different electric motors supplied with constant voltage from SE11: constant, series and parallel excitation. Parasites are added, I14, R13, I24 and R23 for armature inductance, armature resistance, rotor inertia and rotor friction.

separate source, but usually it is supplied from the same power source as the armature. So we have 3 possible circuits:

1. Constant excitation, normally from a permanent magnet.
2. Field coil in series with armature so that both currents are equal: series excited machine.
3. Field coil in parallel with the armature: parallel excited machine.

Box 5.1 contains the equations of these three electric motor circuits neglecting parasites. The following observations apply:

- With constant excitation the machine acts as a gyrator.
- With series excitation torque at constant voltage, torque is proportional to the square of electric current. Torque and power increase strongly at low shaft frequencies. Such electric machines are suitable for vehicles where they can drive as a motor and brake as a generator to supply electric energy back to the source.
- With parallel excitation the field current is determined by the field resistor Rf (which is generally much higher than in the series case); the torque is proportional to armature current. In the equation for voltage and shaft frequency there is the special point that voltage drops out and an equation for shaft frequency remains which is independent of torque and even voltage. So a parallel excited machine runs at a constant frequency and acts as a flow source in the sense of Bondgraphs.
- The equations in the lower part of the box demonstrate the constant shaft frequency: The voltage acting on the modulated gyrator leads to a shaft frequency independent of torque and supply voltage; a Bondgraph is easy to write.

Box 5.1 Electric DC motor equations with coil excitation

$$\left.\begin{aligned} M &= Ki_f i_a \\ u &= Ki_f \omega \end{aligned}\right\}\quad \text{equations of modulated gyrator}$$

Constant Excitation

set $Ki_f = K_c;\ i_a = i$

$$\left.\begin{aligned} M &= K_c i \\ u &= K_c \omega \end{aligned}\right\}\quad \text{fixed gyrator GY}$$

Series Excitation $i_f = i_a = i$

$$\left.\begin{aligned} M &= Ki^2 \\ u &= Ki\omega \end{aligned}\right\}\quad \text{self modulated gyrator MGY modulated by i}$$

$M = u^2/(K\omega^2)$ Torque as function of voltage u, rot frequency ω

$\dot{E} = M\omega = u^2/(K\omega^2)$ corresponding power

Parallel Excitation $i = u/R_f$ R_f = resistance of field coil

$$\left.\begin{aligned} M &= Kui_a/R_f \\ u &= Ku\omega/R_f \end{aligned}\right\}\quad \text{from where follows } \omega = R_f/K \text{ constant rot frequency}$$

Box 5.1 Equations for electric DC machines with different excitations and neglected parasites. With constant excitation, current is proportional to torque and voltage proportional to shaft frequency. With series excitation, torque and power increase at low shaft frequency. Parallel excitation leads to constant shaft frequency corresponding to the Bondgraph with effort source and modulated gyrator below.

In order to pass to simulation, Fig. 5.4 contains Bondgraphs for three cases of excitation (top, center, bottom) with added parasites: Armature inductance and resistance, shaft inertia and friction (I14, R13, I24, R23, respectively). Note that the causality of the modulated gyrator is different from Box 5.1. We thus expect that the characteristics will be influenced by the parasites. As another observation, an I-element can be slipped over a gyrator to become a C-element (sec. 6.4). We thus expect oscillations and resonances with the other I-element. Damping comes from the R-elements as usual; both are confirmed by simulation.

Figure 5.5 contains the simulation of a series excited electric machine of about 12 kW rated power with start from rest with a suddenly applied voltage. No external load torque (M29) is assumed, so the electromagnetic torque M21 is essentially needed to accelerate the rotor (:I24) and to overcome its friction (R23). Of course when the load consists of a simple inertia and resistance, we can add it to the parameters of the above elements. The sudden switch on of the voltage produces a large but tolerable current and torque peak at start.

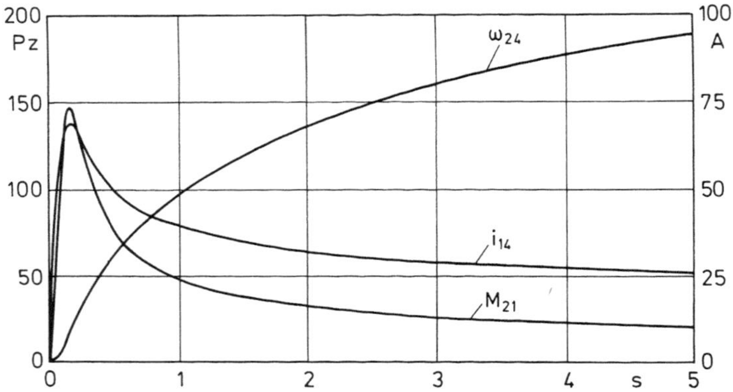


```
Timing:    0.0100000  ,DELTA  ;     5.0100        ,RANGE
PlotBlocks and Scales:
Format:
        BlockNo,   Plot-MINimum,   Plot-MAXimum;   Comment
Horz:        0 ,    0.0000      ,     5.0000     ; Time
  Y1:       24 ,    0.0000      ,   200.0000     ; mech mome o inertia
  Y2:       14 ,    0.0000      ,   100.0000     ; armature inductance
  Y3:       21 ,    0.0000      ,   200.0000     ; eletromagne torque
  Y4:       29 ,    0.0000      ,   200.0000     ; load torque

   0.0000            1 CON                              ;DC series motor
 220.0000           11 SE     111                       ;voltage source
   0.8750000        13 R       14                       ;armature resistance
   0.1750000        14 I       11      -13      -19     ;armature inductance
                              -33
   0.0000
                    19 MGY     24       36              ;armature emf
                    21 MGY     14       36              ;eletromagne torque
   0.0660000        23 R       24                       ;mech friction
   0.8000000        24 I       21      -23      -29     ;mech mome o inertia
   0.0000
   0.0000           29 CON                              ;load torque
   1.7500           33 R       14                       ;field resistance
   0.0307000        36 GAI     14                       ;field torque constant
   1.0000          111 CON                              ;file bonsi52a.sim
```

Figure 5.5 Simulation of a series excited electric motor with sudden application of 220 volt at start up. There are high currents and torque peaks because the electromagnetic force u19 is missing with stopped shaft.. As shaft frequency increases current and torque decrease.

Figure 5.6 shows the departure of a parallel excited electric machine with the same parameters: The initial current and torque are now much larger and out of range. Consequently, the shaft frequency $\omega 24$ stabilizes sooner, after about 2 seconds. After 2.5 seconds a load torque M29 100 Nm is applied, whence electromagnetic torque and current increase. Also shaft frequency $\omega 24$ decreases in distinction to the equations of Box 5.1, where it should be constant: This is the effect of the parasites.

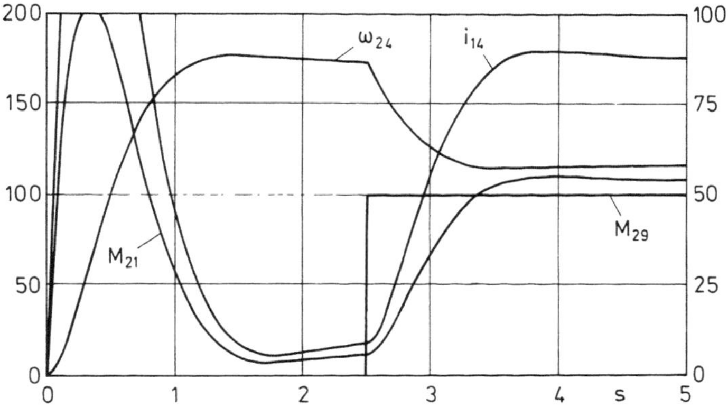


```
Timing:    0.0100000   ,DELTA   ;    5.0100       ,RANGE
PlotBlocks and Scales:
Format:
     BlockNo,   Plot-MINimum,    Plot-MAXimum;   Comment
Horz:      0  ,   0.0000       ,    5.0000       ; Time
  Y1:     24  ,   0.0000       ,  200.0000       ; mech mome o o inertia
  Y2:     14  ,   0.0000       ,  100.0000       ; armature inductance
  Y3:     21  ,   0.0000       ,  200.0000       ; electromagne torque
  Y4:     29  ,   0.0000       ,  200.0000       ; load torque

   0.0000           1 CON                           ;DC parallel motor
 220.0000          11 SE       111                  ;voltage source
   0.8750000       13 R         14                  ;armature resistance
   0.1750000       14 I         11     -13     -19  ;armature inductance
   0.0000
                   19 MGY       24      36          ;armature emf
                   21 MGY       14      36          ;electromagne torque
   0.0660000       23 R         24                  ;mech friction
   0.8000000       24 I         21     -23     -29  ;mech mome o o inertia
   0.0000
   2.5000          29 PLS                           ;load torque
  10.0000
 100.0000
   0.1820000       33 R         11                  ;field resistance
   0.0307000       36 GAI       33                  ;field torque constant
   1.0000         111 CON                           ;file bonsi52b.sim
```

Figure 5.6 Start up of a parallel excited motor. Here torque and armature current peak are much higher but decrease rapidly. After 2.5 seconds a load torque M29 is applied, thus current increases and shaft frequency decreases.

Since the Bondgraph gives only the control characteristics, the following technological observations are of interest:

- The power limit of electric motors comes from the capability to evacuate the heat (entropy) generated in the resistors, particularly R13, without dangerous over temperatures. Since there is a thermal delay (time constant) short overloads are possible (up to 10 times rated torque).

- The armature current has an instantaneous limit (coming from the commutator) which limits the available torque.
- Generally current and torque pulses as simulated are undesirable (f.i. for the comfort of passengers in electric vehicles), therefore, voltage is normally applied gradually, often by inserting electric resistors between source and motor. Such torque peaks due to changing resistors can be easily felt by a passenger, either on Hankyu Railway around Osaka or on the electric street cars of Toronto.
- Sometimes excitation comes from both series and parallel field coils for special characteristics. This is the so-called compound excited electric machine.

5.2.1 AC Induction and Synchronous Motors

Alternating current (AC) electric machines comprise two interesting types:

- Asynchro (Asynchronous) or induction type.
- Synchro (Synchronous) type.

As already treated in Figs. 3.15 and 3.16, asynchronous electric machines are a kind of non-linear clutch between electric frequency and shaft frequency of rotation. They require a slip to produce torque. Synchro machines run at constant shaft frequency and are coupled by a non-linear C-element, which represents a magnetic spring between electric frequency and shaft frequency.

Figure 5.7 contains at the top a Bondgraph of a synchro electric motor. Compared to Fig. 3.15, the coupling resistor is replaced by a capacitor. The torque depends on the torque angle, that is the angle between the phase of the electric current and of the mechanical shaft. It is proportional to the sinus of the torque angle, as calculated in Fig. 5.7 by C141 and by the block 14. Hence:

- With small torque angles, the sinus can be approximated by the angle itself, expressed in radiants. Then the torque is proportional to the torque angle. For practical simulation it can be estimated by observing that the rated torque (rated power divided by shaft frequency) is obtained with a torque angle of about one half radiant.
- With larger load torques the torque angle can go over the peak of the sine curve, the machine slips and falls out of step. With continuous slip it produces no torque at all because the sinus curve has inverse slope, hence torque reverses after the peak.
- The magnetic spring C14 can enter resonance with the rotor inertia I23. Therefore damper windings that increase R25 are sometimes used. These damper windings act like an electric Lanchester damper (Fig. 3.11).

From the above it follows that synchro motors cannot start up from rest: They need a start up motor like an automobile engine. Sometimes a real machine has an asynchro induction winding for start; it is switched off after reaching synchro shaft frequency.

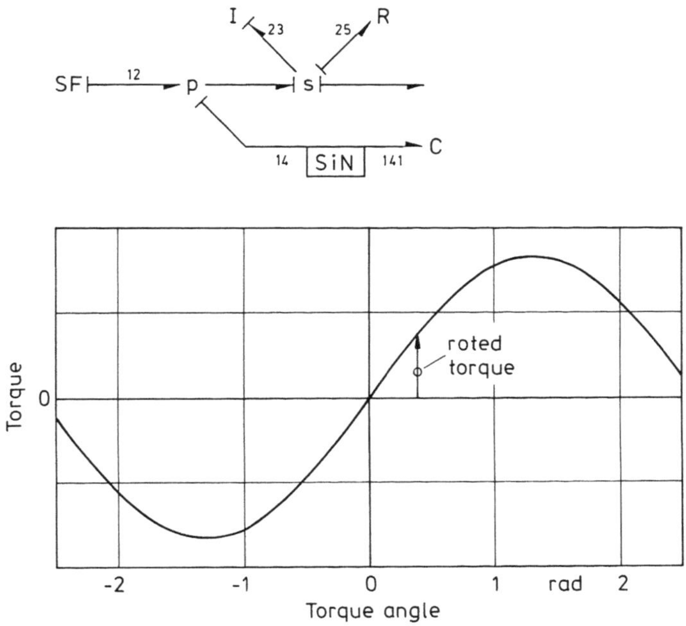

Figure 5.7 Bondgraph of a synchro electric machine for AC. Here torque is controlled by C14 (C141, block 14) and is proportional to the sinus of the angle between electric phase and shaft angle. Oscillations can occur with I23 and are damped by R25. Below: Relation between torque and load angle as a sinus curve.

Turning to the electric side, without torque a synchro machine only draws a small magnetization current which is 90 degrees out of phase with supply voltage. Hence no real power is carried. With required torque the current will increase and become more in phase with voltage (angle nearer to 90 degrees) and draw real electric power. For more on electric AC power see sec. 8.5.

5.3 Capacitors, Inductors, Actuators and Real Transformers

Here we shall describe the action of inductors more closely, including devices based on inductors like magnetic actuators and electric transformers with parasites. However, we shall start with electric capacitors with moving plates, which are conceptually simpler.

Figure 5.8 contains at the top a schematic of an electric capacitor where the plates can have a variable distance h. This device is used in basic electronic engineering to calculate the attraction force between the plates by consideration of energy. More precisely it should be the co-energy if the plates are held at constant voltage.

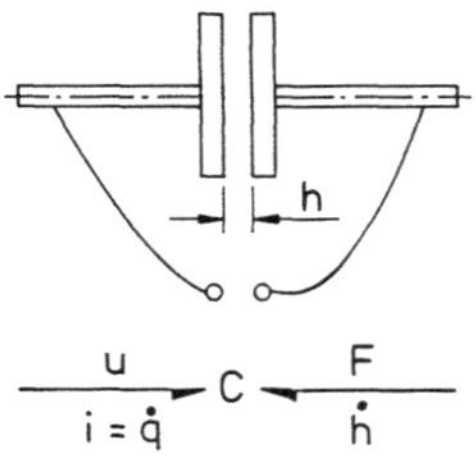

$$\frac{dU}{dt} = u\dot{q} + F\dot{h}$$

$$q = Cu = \frac{\varepsilon A}{h} u$$

$$C = \frac{\varepsilon A}{h}$$

$$F = \frac{q^2}{2\varepsilon A} = \frac{u^2 \varepsilon A}{2h^2}$$

Figure 5.8 Electric capacitor with movable plates, voltage u, charge q, plate separation h and attractive force F. Schematic at left, Bondgraph at right and equations below. C is capacitance, A the plate separation and ε the permeability constant.

It can be seen that an electric capacitor is essentially a 2-port C-field from the fact that capacity depends on the plate separation h. The equations appear below on Fig. 5.8, including the attractive force.

We know from sec. 2.6 that C-fields should have Maxwell symmetry, but only in the linear case. So we expect linear constitutive equations between voltage and force on one side and electric charge and displacement on the other side. Indeed, linearizing the equations of Fig. 5.8 around a point of operation Maxwell symmetry appears, as had been shown in Box 2.2.

In general, attractive force depends on whether a capacitor is at constant charge (wires disconnected) or at constant voltage (wires connected to a voltage source). In Bondgraph terminology, it depends on the causality of the C-field in the system, integral (maintained charge and distance) or derivative (maintained voltage and force). Causality with force input is physically not realistic, because the attraction force would collapse the plates. Only with a linear dielectric (relation between electric D-field and E-field) does the attractive force happen to be equal for both connections or causalities. In particular, the so-called dielectric saturation increases attractive force.

It should be noted that the force is always attractive; this corresponds to the power orientation of Fig. 5.8. In fact, with attraction force and increasing separation taken positive, power flows into the capacitor (strictly into the electric field between the plates) when the plates are separated against the attraction force.

A moving plate capacitor may be used for energy conversion by coupling one plate to a mechanical crank which is a modulated transformer as shown in Fig. 5.9. Here the parameter of the transformer depends on the shaft angle, taken by an integrator block from the rotary side at the right. There is also a switch that couples the plates electrically to a high voltage source when the plates approach, so that the force supplies mechanical power. At minimum distance, the switch couples the plates to a low voltage source, so that during the following separation they experience less force and draw less energy. Consequently, a net mechanical energy remains after one cycle or revolution of a crankshaft. In order to avoid a short circuit with excessive currents and losses when switching sources, there are intermediate phases placed with the plates disconnected (coupled to zero flow junction SF_0). This allows to adjust the voltages before reconnecting the plates to the voltage sources.

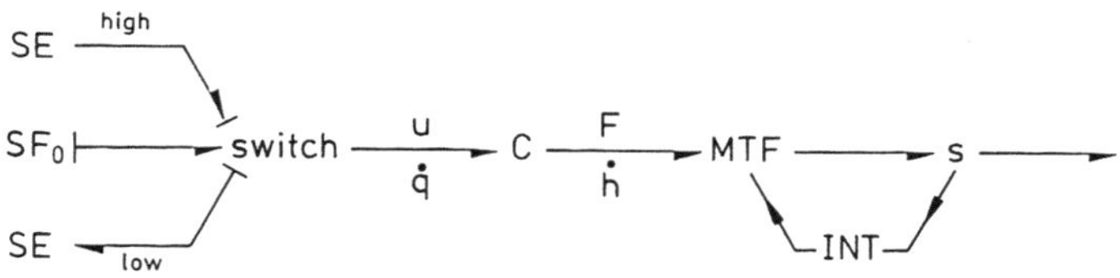

Figure 5.9 Electromechanical conversion engine with a moving plate capacitor and a crank shaft allowing periodic motion of the plates; this shaft is reticulated by an MTF, modulated by the shaft angle. The electrical bond is coupled to the high voltage source when plates approach and to the low voltage source when they separate. In between they are connected to a zero flow source which maintains the charge constant.

The conversion engine of Fig. 5.10 is not very practical but shows the fundamental working of heat engines (where entropy and temperature replace electric charge and voltage). In fact, the intermediate phases were invented for thermal conversion (heat) engines by Sadi Carnot as long ago as 1824. Thus ths operation is called the Carnot cycle. Neglecting parasites it is completely reversible; the engine of Fig. 5.10 can also pump electric charge from low to high voltages thus converting mechanical into electrical power. This is occasionally used as the van de Graaff generator for high voltage.

5.3.1 Electrostatic Loudspeaker or Condenser Microphone

Figure 5.10 shows another application of the moving plate capacitor. Here one plate is built as a soft membrane that moves under the attraction force. Its spring effect is part of the 2-port C-field and its mass becomes a separate I-element. The electric leakage between the plates appears as parallel insulation resistance. Depending on the power direction, the device is a loudspeaker, as indicated, or a microphone. The membrane moves an airflow through a transformer with the approximate membrane area as parameter. It is loaded by the radiation resistance which represents the power radiated into the air and by its mass (really inductance) moving with the membrane. According to acoustic theory, the air inductance, the radiation resistance and the effective membrane area depend on frequency. More precisely these parameters depend on the ratio of acoustic wavelength to the diameter of the membrane.

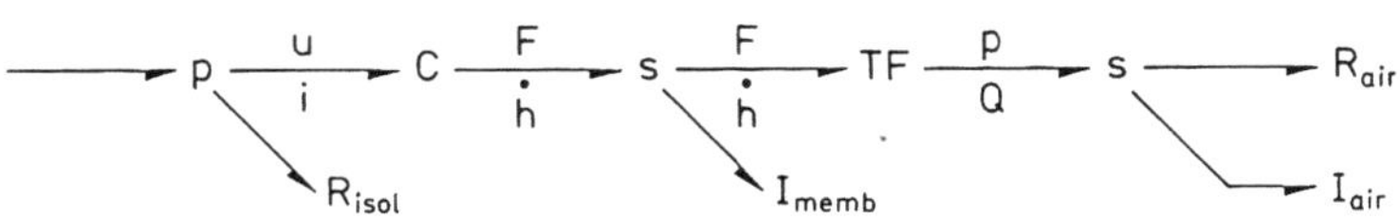

Figure 5.10 An electromechanical C-field followed by a transformer represents an electrostatic loudspeaker and or a condenser microphone, depending on the direction of power. The power arrows on the figure indicate loudspeaker operation. There is the insulation resistance in parallel, also the mass of the membrane. At the right we find radiation resistance and inductance of the air; both depend on frequency.

The Bondgraph of Fig. 5.10 represents acoustic radiation theory correctly, but contains a causal conflict. For simulation it would be necessary to reflect the membrane inertia through the transformer, or to add a C-element for compressibility at some place. The main purpose of the figure is to display the fundamental operation of electrostatic loudspeakers and condenser microphones as these devices are called. They are reversible except for the R-elements. The beauty of the figure is that the Bondgraph represents effects of 3 engineering disciplines.

– Electronics, Mechanics and Hydraulics.

5.3.2 Magnetic Inductors

In an inductor we have a magnetomotoric force (MMF) proportional to electric current and a magnetic displacement derivative (incorrectly called magnetic flux) proportional to induced voltage, as follows from basic electronic theory. This is the action of a gyrator shown in Fig. 5.11, bottom. It changes induced voltage into a time derivative of magnetic displacement ϕ and electric current into the MMF. The magnetic circuit has displacement proportional to MMF, is thus a C-element in itself.

The upper part of Fig. 5.11 contains an inductor with a magnetic circuit in the ring or iron core; the core is interrupted by a variable air gap, controlled by a rod from the right. With soft iron (high permeability) all the energy is concentrated in the airgap and we have the same constitutive equation as with the capacitor of Fig. 5.8. Only magnetic permeability replaces the dielectric constant. Hence, we have the same expression for the attractive force. The gyrator is power conserving and the C-fields energy storing as always.

The final magnetization curve is generated in the C-element, where the B field represents displacement ϕ and the H field the MMF (MagnetoMotoric Force). The curve goes through the origin with soft iron as used in electromagnets, but with permanent magnets we have a displacement without MMF. The parameter of the gyrator is the number of windings (or strictly this number divided by Fleischmann's Constant for correct physical dimensions (Fleischmann 1980)). The magnetomotoric force is also called ampere turns.

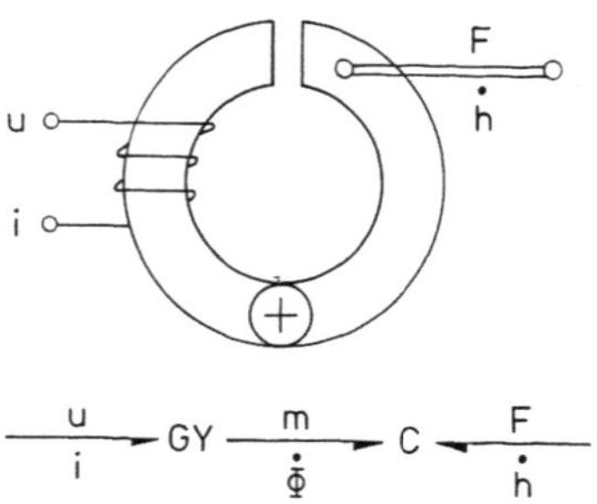

Figure 5.11 Magnetic actuator with a coil and a magnetic circuit, built up of two movable irons with a variable air gap. It acts like a moving plate capacitor between magnetomotoric force m and magnetic displacement ϕ on the magnetic side, and between force and gap separation h on the mechanical side. The gyrator generates the magnetomotoric force and the displacement derivative from electric current and voltage.

As already mentioned, the quantity ϕ is usually called magnetic flux, although it is a displacement in the sense of Bondgraphs that can be maintained in a C-element or C-field without supplying external power. If it were a flux or flow, it had to come from an R-element and require continuous power dissipation. In such a case a permanent magnet could not exist.

5.3.3 Real Electric Transformers

As shown in Fig. 5.12 an electric transformer is really a magnetic circuit, similar to Fig. 5.11 without an air gap and equipped with 2 windings. So the magnetic displacement travels around the outer rectangle through both windings. Only a small part passes through a vertical central leg and avoids the other windings: It is called the undesirable stray displacement (stray flux). The lower part of Fig. 5.12 contains a Bondgraph with a 2-port C-field and 2 exciting gyrators.

The C-field is reticulated in Fig. 5.13, top, as a series capacitor for the coupled displacement and 2 small capacitors for the stray displacement behind p-junctions. Resistances or R-elements for the windings are added. In the lower part of the figure, the left gyrator representing winding 1 is slipped over the

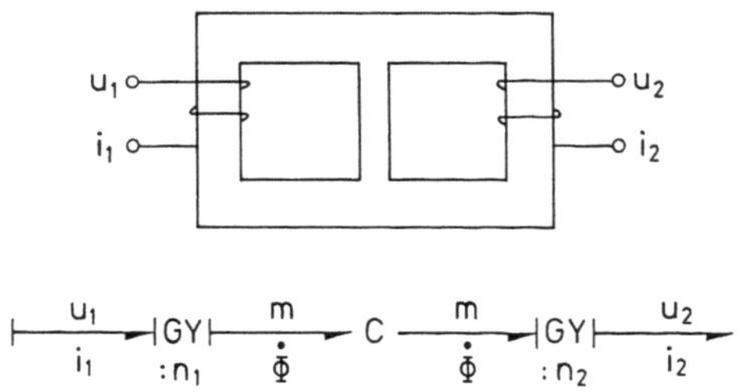

Figure 5.12 A real transformer is like an inductor with 2 windings. There are gyrators attacking a 2-port C-field. The schematic shows that a small part of the magnetic displacement is not coupled to both gyrators, but bypasses or strays through the central iron leg.

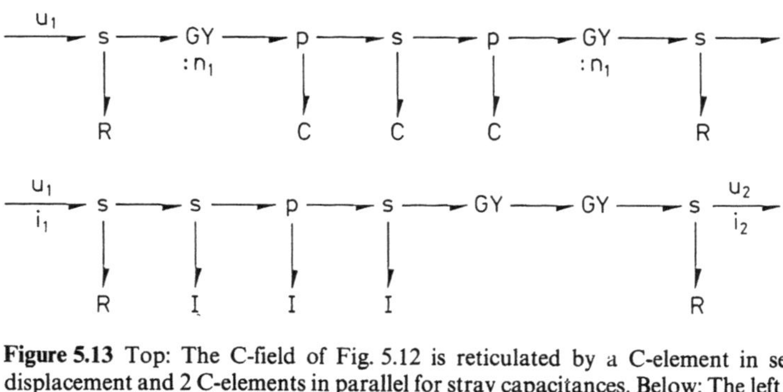

Figure 5.13 Top: The C-field of Fig. 5.12 is reticulated by a C-element in series for a coupling displacement and 2 C-elements in parallel for stray capacitances. Below: The left gyrator is slipped to the right, whence the C-elements become I-elements and the junctions change from s to p and from p to s. The coupling inductance at left now represents a short circuit for DC that prevents electric transformers working. The 2 gyrators in cascade act like an ideal transformer.

junctions to the right. As we shall see in sec. 6.4, this changes p-junctions into s-junctions and vice versa, further the C-elements into I-elements. Both gyrators are now cascaded at the right and act like an ideal transformer.

The parasites in Fig. 5.13 are the inductors in series (stray inductors) and in parallel (coupling inductors), in addition to the winding resistances. With very low frequency, the impedance (complex resistance) of all inductors decreases until with DC (direct current) it becomes zero. Then the coupling inductor represents a short circuit (compare sec. 6.3) and this is the reason why real electric transformers cannot work with DC.

There are still some points to note:

1. We have not applied causalities to Fig. 5.13 since the capacitor or inductor network would give causal conflicts. Simulation can be done without difficulty from the 2-port C-field of Fig. 5.11, using the former Fig. 2.33.
2. Introducing a small air gap into the magnetic circuit (iron core) decreases the coupling inductance and increases magnetic stray displacement.
3. If the iron has hysteresis, it could be added to the capacitors by additional blocks.
4. The iron core and the windings experience mechanical forces, similarly as treated with Fig. 5.11. This leads to the known and objectionable noise of large electric transformers.

Chapter 9 contains a worked example of simulation of magnetic hysteresis.

6 Computational Overview, Practical Procedures and Problems

6.1 Programs for Simulation from a Bondgraph

Most worked examples in this book are done from TUTSIM, because it is powerful and fast. However, Bondgraphing is universal and not limited to one or the other program. In essence, a fully augmented Bondgraph (with all power directions and causalities) is equivalent to a set of equations that can be programed directly in language like BASIC, PASCAL, FORTRAN or C.

In practice it is usually time saving to use an application program; we mention here the programs MICRO ENPORT and the couple CAMP/ACSL in addition to TUTSIM.

Figure 6.1 shows a map of simulation strategy. We have to add the parameters to the Bondgraph system model and then to use one of the programs to arrive at the interpretation below.

Some observations to Fig. 6.1:

MICRO ENPORT has been adapted from ENPORT (1988), a program originally conceived for big computers. Version 7 was released in 1988 for PC (IBM standard) and also allows non-linear state equations. Further, it calculates the Eigenvalues.

TUTSIM (1988) "Technical University Twente Simulation" was originally conceived for simulation from a blockdiagram and written in assembler by Vim Meermann. Jan van Dixhorn recognized the possibility to input directly from a Bondgraph. It is a block oriented language with many powerful blocks, but gives only time domain results, no frequency response or Eigenvalues.

FANSIM (1988) was developed by Ken Fyfe under the guidance of Jean Thoma. It started as a postprocessor for TUTSIM, reading the TUTSIM numerical output from a diskette, applying FFT (Fast Fourier Transform) to obtain graphical results in the frequency domain. The version 2.1 (spring 1988) has many added features for modal analysis, like frequency response, poles and complex residues and finally transfer matrices.

CAMP (1988) is a preprocessor for ACSL (1988), a large powerful and widely used simulation package. CAMP writes the structure code with a similar input as ENPORT but without parameters; they have to be added after transfer to ACSL. This last one can linearize non-linear equations and calculate Eigenvalues, a feature particularly useful in a Bondgraph context. Originally conceived for

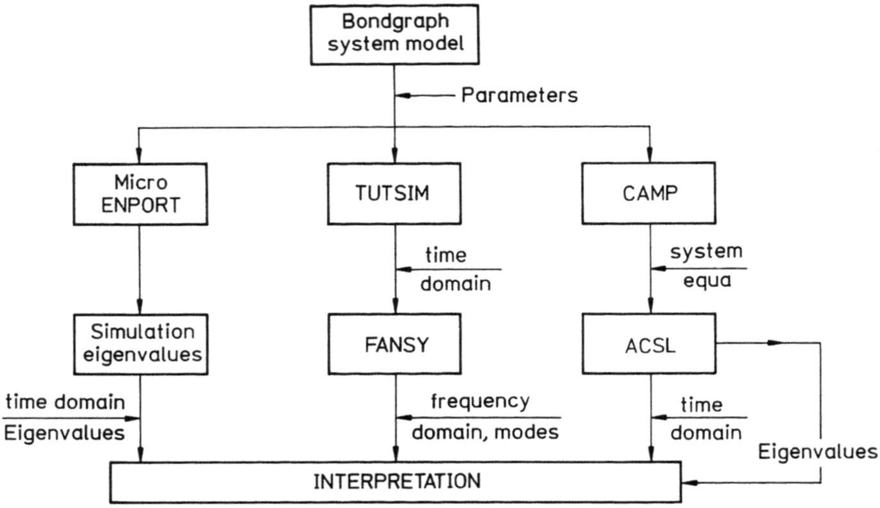

Figure 6.1 Map of simulation strategy; from the Bondgraph model with parameters we arrive at results by different ways or programs. An alternative would be to write equations directly and to solve them in a higher language such as PASCAL.

minicomputers, the PC version is still (spring 1990) somewhat slow. The author cannot prevent himself from thinking that it would be much faster if it were rewritten in a modern language, preferably C, instead of Fortran. As it is, many people program ACSL from a Bondgraph without going through CAMP, but in the experience of the author's group this is more tedious.

Eigenvalues can also be obtained by writing the state equations by hand from a (manually) linearized Bondgraph, a good and simple student exercise. To calculate the Eigenvalues from the state equations, any manual approach is prohibitive for more than 3 state variables, but a package like MATLAB (1988) does this very well. Obviously, it would be desirable to implement the entire process on a computer, preferably from a TUTSIM listing, a task that seems to be under way.

6.2 Linear Control Theory, Eigenvalues and Calculating Step

We shall summarize linear control theory here as far as necessary for efficient use of the computer. For more information the reader is referred to the text of Takahashi (1970). Such linear theory is also applicable to non-linear Bondgraphs after linearizing (Taylor expansion around a point of operation).

Bondgraph computation is mostly done with state variables and state space equations. They are a set of coupled differential equations of the first order with time independent coefficients. It is convenient to write them in the matrix-vector

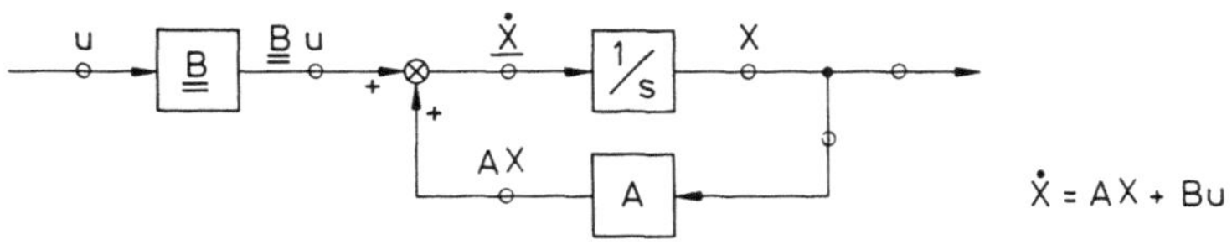

Figure 6.2 State equations and multi-blockdiagram. Such equations feature in most automatic control theory. Often there is another line added with C and D matrices but is not necessary in this book.

form of Fig. 6.2 (right), illustrated by a simple blockdiagram at the left. The equations in Fig. 6.2 are called canonical state space equations because of their importance. The quantities U and X are input and state vectors (in the sense of matrix algebra) and A and B are matrices.

State variables are said to fix the state of a system, a somewhat circular definition that does not help much to distinguish them from other variables. We like to relate them to the degrees of freedom of a system: each such degree can be controlled and freely fixed by the value of a variable, which is then a state variable. Other variables follow by internal constraints (equations). A complete set of state variables controls all degrees of freedom of a system. Naturally there is some latitude in the selection of state and other variables. In the context of simulation,

Box 6.1 A – Matrices and Eigenvalues.

The Eigenvalues of an n∗n matrix are the solutions of

$$\text{Det}\,|A-\lambda|=0$$

giving $\lambda_1, \lambda_2 \cdots \lambda_n$ Eigenvalues (EV)
EV are either real or conjugate complex (in pairs)

$$EV = RE + j\,Im \text{ (Real, Imaginary parts)}$$

System modes $x = x_0 \exp[Re + jIm)t]$
Overdamped case $Im = 0 = 0$, $R_e = -1/r$

$$x = x_0 \exp(-t/\tau)$$

underdamped case $Re + Im = -\xi\omega_n + j\omega_n(i - \xi^3)^{0.5}$
where $\omega = (Re^2 + Im^2)^{0.5}$, damped resonance frequency

$$\zeta = -Re/\omega_n \text{ damping}$$

Notes:–EV have dimensions of time and are expressed in Rad/Sec
–For weak damping, $\xi \ll 1$, say < 0.2, we have
$\omega = Im =$ undamped resonance frequency

Box. 6.1 Relations of A matrix, EV, resonance frequency, damping and time constants.

each state variable can be given an independent initial condition. The evolution follows then from the state space equations.

Matrix B shows how the input signal is coupled into the system and matrix A controls the system dynamics. The state vector has as many components as necessary to determine the state of the system, in general as many as there are integrators. Further, a system with n integrators produces a state vector of n components and an A matrix of n * n (n times n) size. As developed in Box 6.1, a set of n numbers called Eigenvalues (abbreviated EV) can be calculated from each A matrix. They are really frequencies or reciprocal times and control the dynamics of the system, which is stable if all EV have negative real parts. The imaginary part controls the frequency of the oscillations, the real part is the damping. In case of real EV, there will be no oscillation but only smooth decay of an initial disturbance.

6.2.1 Bondgraphs and State Space Equations

The important point is that all linear Bondgraphs with integral causality are equivalent to the state space equations of Fig. 6.2. Simulation can be done by intergrating these equations, a standard task of numerical mathematics. Each state variable is the output of an integrator and for the Bondgraph the best choice is:

- displacements (q-variables) of the C-elements
- momenta (p-variables) of the I-elements

to which should be added the output of any signal integrator if uncoupled to the p and q just mentioned).

Consequently the number of C- and I-elements give directly the number of state variables or, as called in higher control theory, the numbers of dimensions of the state space.

The following points are of interest:

1. The state variables are coupled only by the canonical equations in Fig. 6.2 and statically independent or uncoupled. This allows to give them independent initial conditions.
2. If a Bondgraph contains a derivative causality, the displacements or momenta are no longer statically independent and the canonical equations can no longer be written. Hence, one of them is not a true state variable and should be eliminated.
3. The EV control the necessary time step for computation. If they are widely separated, say by a factor of 100 or more, we have a stiff system, difficult to compute.

As stated before, writing canonical state space equations by hand from a Bondgraph with only integral causalities is not difficult.

6.2.2 Eigenvalues and Time Step

The selection of step size for computation is important together with the maxi length of simulation. We denote:

dt = time step; tmx = maxi length of simulation

The ratio of maxi length to time step gives the number of calculating steps. More steps require more computer time and can lead to accumulation of rounding errors. The author has a feeling, but no proof, that rounding errors do not accumulate for AC (alternating current) signals. Consequently a good step number is between 1000 and 20000 steps. They will require about 0.5 to 10 minutes run time on a PC-AT type of computer with a Bondgraph of 50 bonds.

Increasing the time step allows to simulate a longer time interval with a given number of points. The maximum value of the time step is:

- 4 steps per reciprocal of the highest EV.
- 4 steps per radiant of oscillation (25 steps per period).
- 4 steps per time constant.

The underlying mathematical theory is complex (Gear 1976) and leads to twice the above step size. However, such a larger step has led the author to grossly inaccurate results and even divergences.

According to these rules, the highest EV, resonance frequency or shortest time constant forces the time step on us. So in a stiff system, the shortest time constant determines the necessary step, and much longer time constants or slow oscillations require many calculating points. In practice, the small C-elements introduced to avoid derivative causality (sec. 6.4) are often troublesome.

EV can be reduced by tricks: Arbitrarily increase some C and I at critical places, as treated in detail with their disadvantages and possible bad conscience in sec. 6.4.

6.3 Stability Preview and Steady State Bondgraphs

The purpose of a Bondgraph simulation is to find all the variables of a real installation, both in steady state and as function of time. Hence variables have both DC (direct current) and AC (alternating current) components of different frequencies. The DC and some lower frequency AC components are normally desirable, but the higher components are nuisances. They represent vibrations which stress the real installation or system without useful output. On a unit step input, damped AC components coming from stable EV lead to vibrations, undamped or negatively damped components indicate instability, which must be precluded by design and simulation.

For a preview of stability and vibrations, it is helpful to look on energy fluxes or power in general without thinking of their composition of efforts and flows;

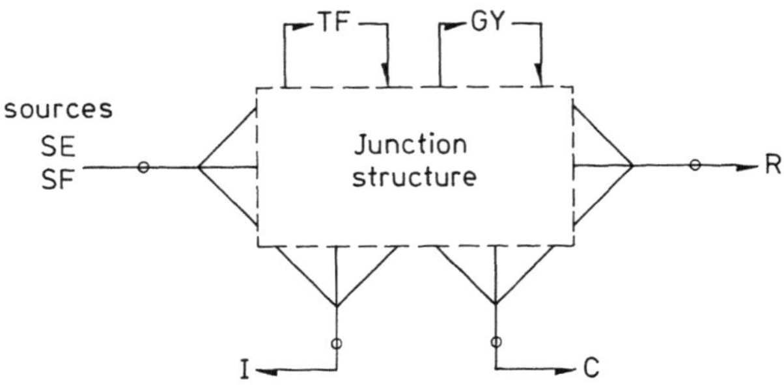

Figure 6.3 Summary Bondgraph with sources on the left, energy dissipators on the right, power conserving elements at the top and energy stores below. In the center there is the possibly complex junction structure. Power passes on average from right to left, that is from the sources to the dissipators. In addition oscillatory modes may be excited and lead to power exchange between the energy storers.

this can be done by inspecting the Bondgraph. As shown in Fig. 6.3, the elements of most Bondgraphs can be arranged according to their energy treatment: Sources SE and SF at the left, power conservers TF and GY at the top, power dissipators R at the right and energies storers C and I below. In the center there is the junction structure, a maze of s- and p-junctions. (This picture is adapted from a similar one in Karnopp and Rosenberg 1975, Fig. 7.25.)

So in a real steady state power is supplied from the source in Fig. 6.3 and runs through the junction structure (called a switch yard for power by Karnopp and Rosenberg 1975) towards the R-elements at the right for dissipation. The generated entropy or heat is excluded here, see chapter 7.

Still in steady state, no power enters or leaves the energy storers C and I, which are filled to a constant level. However, with any change of the inputs transients are set up that can lead to intense power exchange between the energy storers, perceived as free vibrations of the system. With AC sources forced vibrations will be excited; they are called resonances if the exciting frequency is near to a weakly damped EV. The same is true for random excitation.

Vibrations and resonances are generally objectionable since they stress real machine members, that is they impose far greater efforts and flows than needed to conduct the DC power. Consequently, Eigenvalues, especially the higher ones, should have sufficient damping and not be coupled too tightly to the main structure. This is another reason for the wish to know the Eigenvalues in addition to the choice of the time step in sec. 6.2.

In principle EV can be determined by running the matrix through a suitable program package, like the MATLAB (1988) already mentioned. In practice, it is useful to estimate them by inspection of the Bondgraph or by looking at the corresponding electric circuit.

6.3.1 Static Bondgraph for Steady State

We are often interested in estimating the steady state from a given Bondgraph, ready to compare with the final values from a (dynamic) simulation. For this we can simplify a Bondgraph by the following observations, essentially removing the energy stores which are filled to a constant level. They are illustrated in Fig. 6.4.

- C-elements are filled to constant effort and become SF_0 (flow sources with zero flow and any effort).
- I-elements are filled to constant flow and become SE_0 (effort source with zero effort, any flow).

The effect of this change is also developed in Fig. 6.4; it can be summarized;

- For a C-element on a p-junction, its own bond disappears and other bonds remain as they are.
- A C-element on an s-junction makes all adjacent bonds (bonds of the same junction) disappear.
- An I-element on a p-junction makes all adjacent bonds disappear.
- An I-element on an s-junction makes its own bond disappear and leaves the other bonds as they are.

As an example, Fig. 6.5 shows at the top a linear Bondgraph and in the middle a corresponding electric circuit. According to the above rules, all C-elements

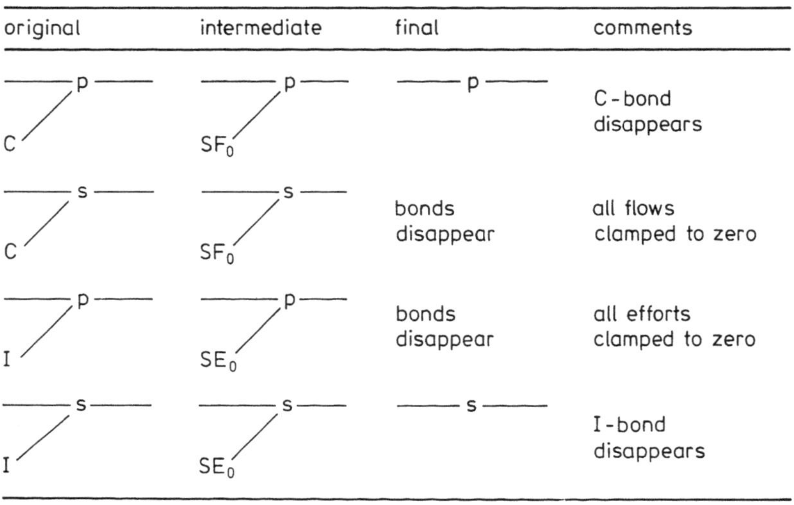

Figure 6.4 For obtaining a steady state Bondgraph the C- and I-elements are filled to constant level and become zero flow source and zero effort sources, respectively. They either disappear or make all bonds on the next junction disappear.

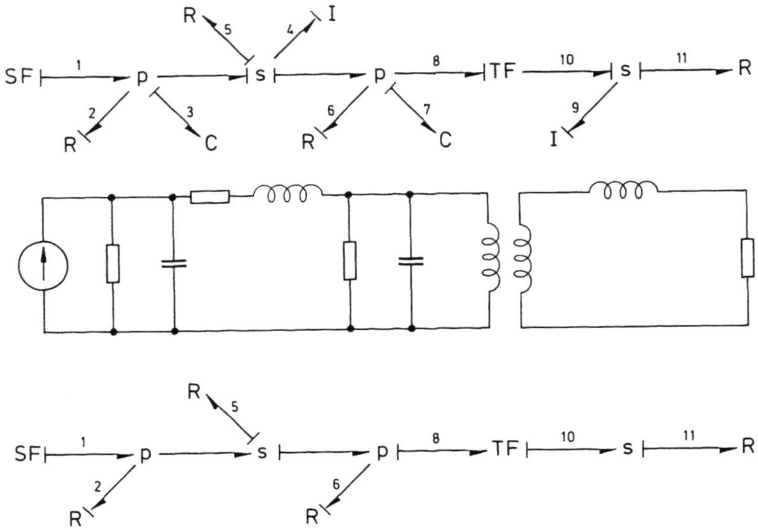

Figure 6.5 Example of a steady state Bondgraph at the bottom, obtained from the full dynamic Bondgraph at the top, with the corresponding electronic circuit in the middle.

disappear because they are on p-junctions, also all I-elements because they are on s-junctions. The remaining Bondgraph below for steady state is much simpler. The values of the variables can easily be calculated by hand or by a suitable circuit program, but definitely not by the programs cited above because it is full of algebraic loops treated in sec. 6.5.

Figure 6.6 contains a similar example, only the central inductance has been replaced by a capacitor C14. It makes all bonds on the central extension disappear by clamping them to the common flow of zero; hence the 2 separate part Bondgraphs below remain. The left part has an effort and a flow determined

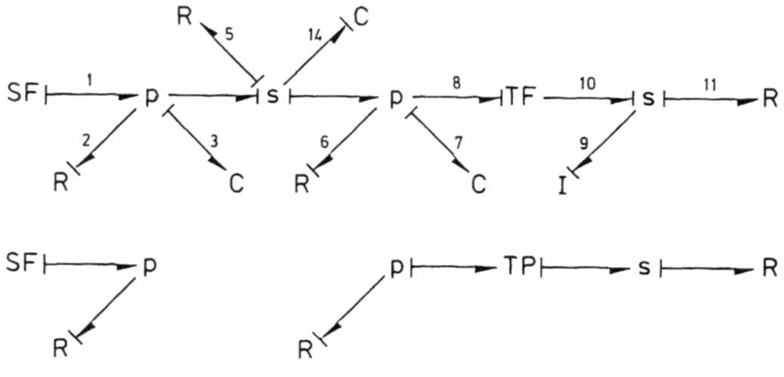

Figure 6.6 Another example of a dynamic Bondgraph at the top, which splits for steady state into 2 part Bondgraphs. The right one has no source and will be motionless.

by R2, but the right one has no source to drive it and will become motionless (neither effort nor flow).

Returning to the full dynamic Bondgraph, the steady state will be reached at constant source setting as the different oscillatory modes decay according to their EV or resonance frequencies and dampings. As soon as the source settings are changed the modes will be excited again.

The speed of decay depends on the real part of the corresponding mode and is generally much faster in high modes. Exceptions are the very weakly damped high frequency modes sometimes occurring in mechanics. So qualitatively we can neglect the higher modes which are outside our range of frequency of interest; very often they are not realistic on simulation because of other high frequency effects not modelled by our Bondgraph. For computation, however, they are very much alive and force a small time step on us (sec. 6.2).

Finally it should be noted that a dynamic simulation is occasionally used even if one is mainly interested in the steady state which appears as the final value: Such is the power of modern microcomputers.

6.4 Eigenvalue Estimation and Dynacheck

The EV estimation presented below is useful in practice but definitely not mathematically regorous: in other words we have 90% probability that frequency and time constants are accurate within 10% and damping within 25%. If we hit one of the exceptional 10%, the case is still worth investigating.

Basically we suppose that in a big Bondgraph nearby elements form subgroups with Eigenvalues that can be determined by inspection. We know that such subgroups are detuned (changed in frequency) by coupling with the remainder of the Bondgraph, but anticipate this to be small. In order to apply the idea, we need to recall the resonance frequency of simple series and parallel circuits, and the technique of reflecting elements through transformers, a well known technique in electric circuit theory. It is called transformer and gyrator slipping in Bondgraph terminology.

6.4.1 Series and Parallel Circuits

Figure 6.7 shows a Bondgraph of a series resonance circuit with common flow as output. It behaves like a second order element, its transfer function is below with a formula for resonance frequency and damping. The damping is proportional to the resistance in series, and the root expression following is necessary to make damping non-dimensional.

Figure 6.8 shows a similar Bondgraph for a parallel circuit with the common effort as output. It also has a transfer function of a second order element with equal resonance frequency and damping proportional to the reciprocal resistance. This should be expected since it is connected in parallel.

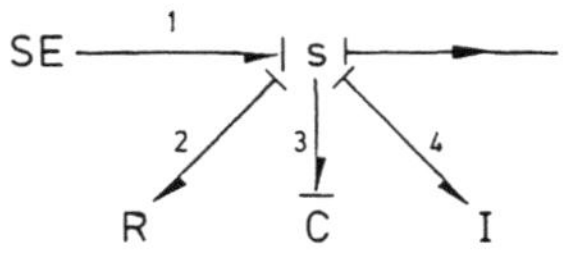

$$e_3 = e_1 - e_2 - e_4 = e_1 - Rf - sIf$$

$$f = sCe_3 = \frac{sCe_1}{1 + s^2 IC + sCR}$$

$$\omega_n^{-2} = IC$$

$$2\varrho = CR\omega_n = R\left(\frac{C}{I}\right)^{0.5}$$

Figure 6.7 Bondgraph of a series resonance circuit with transfer function and resonance frequency with damping below. Damping is proportional to the resistance in series.

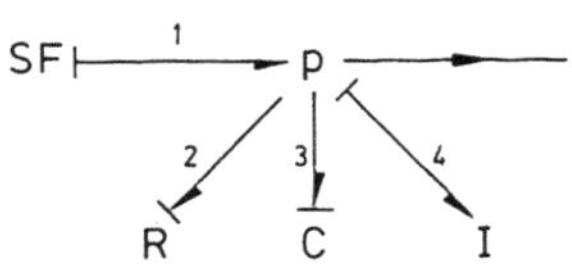

$$f_4 = f_1 - f_2 - f_3 = f_1 - e/R - esC$$

$$e = sI f_4 = \frac{sI f_1}{1 + s^2 IC + sI/R}$$

$$\omega_n^{-2} = IC$$

$$2\varrho = 1/R \times \left(\frac{I}{C}\right)^{0.5}$$

Figure 6.8 Bondgraph of a parallel resonance circuit, again with transfer function and resonance frequency and damping. Here damping is proportional to the reciprocal resistance (conductance) since the resistor is in parallel.

6.4.2 Transformer and Gyrator Slipping

Figure 6.9 shows at the top an I-element behind a transformer and below the same I-element reflected or slipped over the transformer; the Bondgraph is in integral causality. As developed at the right the parameter of the new, reflected I-element equals the original one divided by the square of the transformer ratio.

Figure 6.10 shows the same for the C-element in integral causality. Again the parameter of the slipped C-element is the original one divided by the square of the transformer ratio.

An example would be the hydraulic servomotor of Fig. 2.23. Its transfer function corresponds to the transfer function of Fig. 6.8 if I12 is slipped over the transformer, which has the ratio of A. Details would lead us too far away.

So remains the general rule: If in a resonance circuit an element is separated by a transformer, it has to be slipped over before the formula of Fig. 6.7 or 6.8 can be applied.

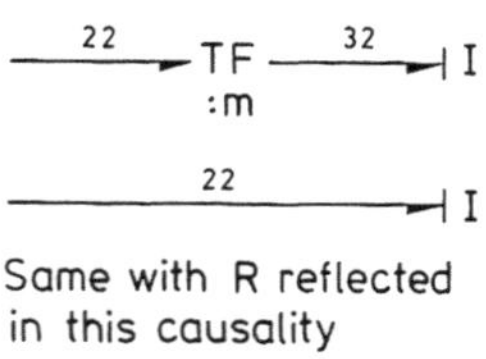

$$f_{22} = m \frac{1}{sI} m$$

$$f_2 = \frac{1}{sI_{ref}}$$

$$I_{ref} = I/m^2$$

Figure 6.9 Reflection of an I-element through a transformer produces another I-element; its parameter is the original parameter divided by the square of the transformer ratio.

Same with conductance
G = 1/R reflected in this
causality

$$e_{21} = m \frac{1}{sC} m f_{21}$$

$$e_{21} = \frac{1}{sC_{ref}} f_{21}$$

$$C_{ref} = C/m^2$$

Figure 6.10 Reflection of a C-element through a transformer giving another C-element. The parameters change as with the I-element.

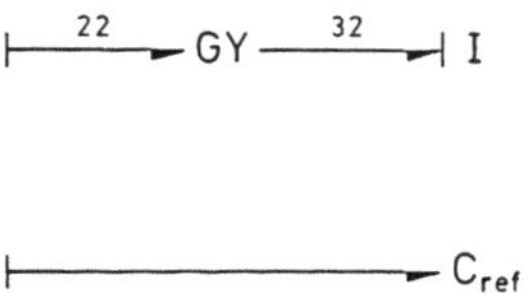

$$f_{32} = \frac{1}{sI} e_{32}$$

$$e_{22} = m \frac{1}{sI} m f_{22}$$

$$e_{22} = \frac{1}{sC_{ref}} f_{22}$$

$$C_{ref} = I/m^2$$

Figure 6.11 Reflection of an I-element through a gyrator produces a C-element. The parameter is similarly divided by the square of the gyrator ratio.

Figure 6.11 contains an example of the slipping of an I-element over a gyrator. It effectively becomes a C-element that can form a resonance circuit with another I-element. An example would be the armature inductance of a DC electric motor that can resonate with the shaft inertia (sec. 5.2). Again with integral causality the parameter of the new C-element is the original one divided by the square of the gyrating ratio. Similarly, if a C-element is slipped over a gyrator, it becomes effectively an I-element.

6.4.3 Slipping and Parameters Changes

The principle of transformer and gyrator slipping is easy, remembering how the parameters change is difficult since the change depends on the causality. Therefore the following rule:

With C- and I-elements in integral causality, the parameter after slipping is the original divided by the square of the transformer or gyrator ratio. The same is true for R-elements in conductance causality: the slipped parameter equals the resistance divided by the square of the transformer or gyrator ratio.

6.4.4 Examples of Eigenvalue Estimation

As stated before, we look on nearby elements and suppose that some Eigenvalues are close to their resonance frequency and to damping. Refinements can be added as in the following example:

Looking back to Fig. 6.5, if I4 and R5 are small they can be neglected. Then C3 can be added to C7, and R6 to R2, leading to the simplified Bondgraph of Fig. 6.12. The new capacitor now has the sum of the former capacitors as parameters since they are in parallel. The same is true for the reciprocal resistance in the new R2.

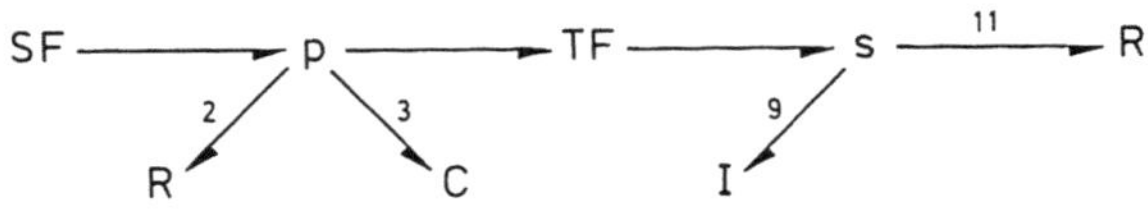

Figure 6.12 Simplified Bondgraph from Fig. 6.5 with central resistance and inductance neglected. It is now a simple second order element and its EV can be estimated according to Fig. 6.8.

Next I9 can be slipped over the transformer and the parallel resonance formula of Fig. 6.8 applied. It gives resonance frequency and damping due to R2.

Alternatively, C3 could be slipped over the transformer in Fig. 6.12 to the right to give a series resonance circuit. Applying the series formula of Fig. 6.7, the same resonance frequency results and some damping due to R11.

In practice, the first pair of EV of the Bondgraph in Fig. 6.5 is very near to the above result, with damping of R2 and R11 added. This illustrates a general property. In a weakly damped system, the total damping is nearly the sum of the dampings caused by each resistor.

6.4.4.1 Resonance of 14

Looking once more at Fig. 6.5 and supposing I9 immobile for the moment, there appears the possibility that I4 resonates through both C3 and C7, which are then effectively in series. This mode leads to the Bondgraph and electric circuit of Fig. 6.12.

The fact that C3 and C7 are in series is apparent from the Bondgraph, because the same flow f4 enters into them and their efforts both act on I4. Thus our formula of Figs. 6.7 and 6.8 can be applied; they lead to another pair of EV with resonance frequency by I4 and one half of C3. The damping is due to R5 and further due to R2 with R6. A person familiar with electronics will easily see the same on the circuit in Fig. 6.12 below.

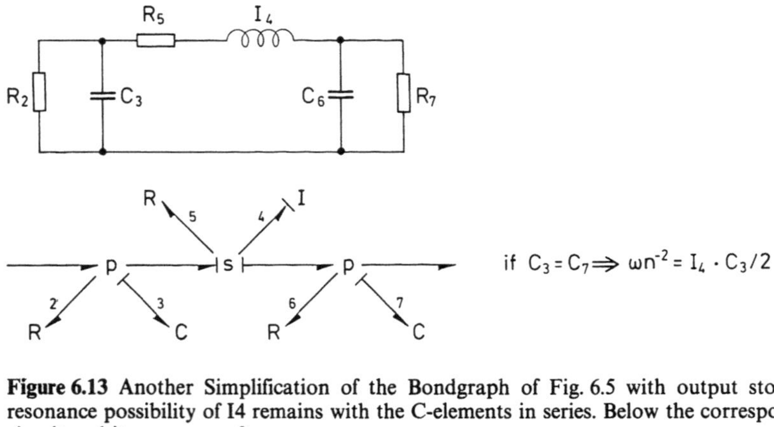

Figure 6.13 Another Simplification of the Bondgraph of Fig. 6.5 with output stopped. Here a resonance possibility of I4 remains with the C-elements in series. Below the corresponding electric circuit and its resonance frequency.

As a computation with an EV package reveals, the above estimation is surprisingly accurate with suitably small R5 and I4; this also justifies to suppose I9 be immobile, as done above. The parameter values come from the simulation of a hydrostatic transmission with the same Bondgraph (sec. 4.3).

We know from the theory (sec. 6.2) that the A matrix of the Bondgraph of Fig. 6.5, which we used here, has only 4 EV since there are 4 energies stores (C3, C7, I4, I9). Therefore no more resonance modes are expected.

The sources SF or SE can be left out in an EV estimation since they are DC whilst we considered AC. Here they become an open circuit (interrupt) for SF and a closed circuit (short circuit) for SE.

6.4.5 Lanchester Damper with Variable Friction

Figure 6.14 contains a Bondgraph of a flywheel I14 coupled by R15 to a series resonance circuit built by I4, R5 and C6. It is similar to the Lanchester damper for torsional oscillations of Fig. 3.11, although an electrical circuit could also easily be written. The mechanical Bondgraph is driven through a gyrator by a flow source; it could represent a classical ammeter, where the gyrator models the change from electric current to magnetic force. Also in such voltmeters the damping of the indicator is an important question.

The problem is to estimate the EV, of which there are 3, since we have 3 energy storing elements (I4, C6 and I14). Supposing the flywheel weakly coupled (R15 small), there is a simple series circuit comprising I4, R5 and C6 with its resonance frequency and damping. This gives the first two EV.

Further we expect a real EV built from the flywheel with a time constant of I14/R15. It is very long if coupling is weak. Thus we have our 3 EV.

However, the situation becomes interesting with tight coupling, whence the time constant goes above the reciprocal resonance frequency and determines the calculation step. It then actually has the value of $\tau = \mathrm{Ired}/\mathrm{R15}$ with the reduced inertia Ired equal to the smaller of I4 and I14, as resulting from the fact that they are effectively in parallel. We can see this also from the Bondgraph where the effort out of R15 attacks both I14 and I4. In the limit, with R15 approaching infinity, the calculation time becomes zero and the Bondgraph incomputable. Infinite resistance in parallel means rigid coupling and in this case one of the I-elements needs derivative causality.

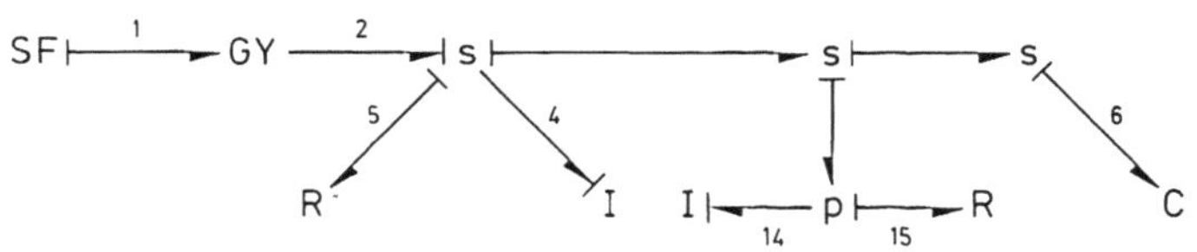

Figure 6.14 Bondgraph of a series circuit, f.i. an ammeter with an I-element I14 coupled by R15. If the coupling is very tight, we get a very high EV, that makes the Bondgraph incomputable in the limit. This is the transition to derivative causality.

The above behavior of Eigenvalues demonstrates a general trend: If we make a coupling element breaking a causality conflict so large (or small in other situations) as to make extremely tight coupling, an EV becomes so large as to make the Bondgraph incomputable. This is the transition to a causal conflict. The remedy is to eliminate R15 altogether and to combine I4 with I14 to avoid derivative causality (sec. 6.5).

With very tight coupling, I14 is practically in series with I4, so the equivalent inertia value is equal to their sum, and we expect the resonance frequency to decrease correspondingly.

Concluding our Eigenvalue estimations, the procedure given is not exact but certainly intuitive. Its practical value is great for determination of the required time step. It is especially useful when the time step needed (to avoid calculation error) turns out to be smaller than expected. A look on the Bondgraph usually quickly reveals the offending parameters and remedies can be applied.

6.5 Causality Problems and Algebraic Loops

We shall treat here two obstacles to efficient computation. It is the great merit of the Bondgraph method that they can be recognized on the Bondgraph and remedies prepared before fighting with the computer.

Mandatory causalities must always be fulfilled because otherwise the Bondgraph is incomputable. Desirable integral causalities on energy storers (C-elements and I-elements) should be obtained by the systematic assignment procedure (sec. 2.3), but cannot be always: we then speak of a causal conflict. Here we describe how the conflict arises on a Bondgraph and discuss some remedies. Mathematically we know from sec. 6.2 that derivative causality arises when state variables are statically coupled (instead of the state equations of Fig. 6.2). Recall that state variables are the displacements of C-elements and the momenta of I-elements.

An example of a causal conflict appeared with a hydraulic actuator on Fig. 4.12; it had 2 C-elements in parallel over a transformer. The remedy was to interpose an I-element for the mass of the casing. We generalize from there.

Whenever 2 C-elements are connected to a p-junction, one of them is in derivative causality as shown in Fig. 6.15 at the top. Their state variables are statically coupled because efforts are equal and statically (algebraically) related to the displacement; in the case of linear C-elements they are proportional as written in Fig. 6.15. The remedy is simple and appears at the right: Just combine both into one equivalent C-element having the sum of the capacities as parameter.

As shown in Fig. 6.15 below, such conflict occurs also when one C-element is separated by a transformer or if both C-elements attack different p-junctions separated by a transformer. The remedy is similar: Slip one C-element (here C24) over the transformer to the p-junction (see sec. 6.4) at left and combine as above.

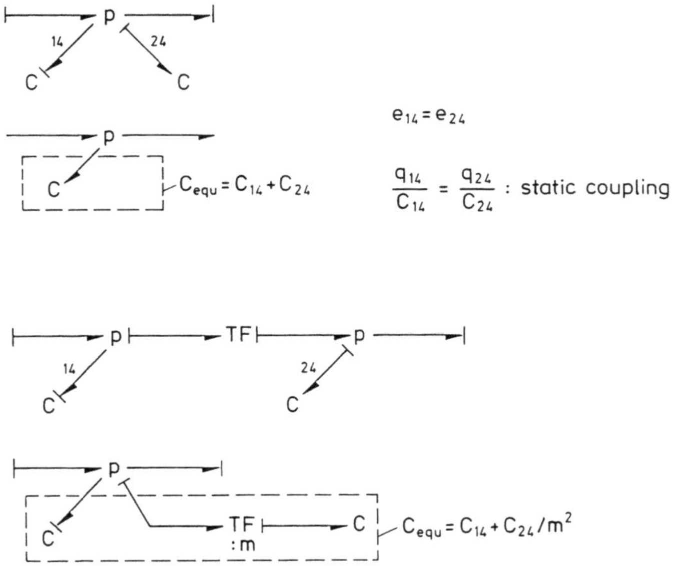

Figure 6.15 2 C-elements on a p-junction lead to a derivative causality since they have static coupling between their state variables. The remedy is to combine them in one capacity. Below: 2C-elements on a p-junction separated by a transformer also give derivative causality. The remedy is to slip one over the transformer and combine it as mentioned above.

No difficulty in principle with a fixed transformer, but the change of parameter involves tedious hand calculation in practice.

Two I-elements attacking the same s-junction or two s-junctions separated by a transformer lead to another derivative causality. The remedy is as above: Combine into one equivalent inductance having the sum of the part inductances (I12, I22) as parameter, perhaps after transformer slipping. Again no fundamental problem with fixed ratio transformers.

Figure 6.16 illustrates a case where transformer slipping is impractical. We have two s-junctions with I-elements separated by a modulated transformer.

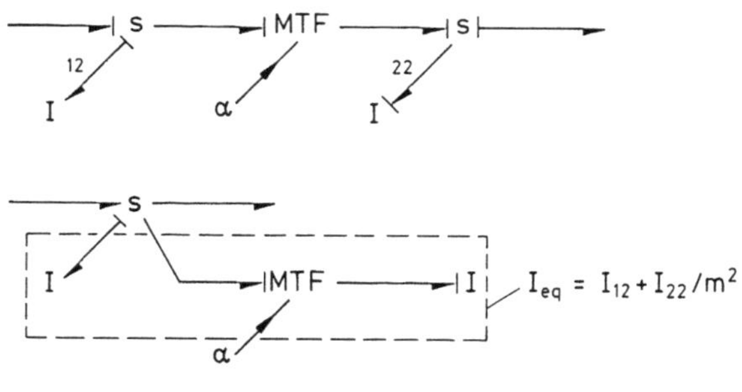

Figure 6.16 I-elements on an s-junction or on 2 s-junctions separated by a transformer give a derivative casuality. The remedy of slipping over the transformer applies in principle, but it is impracticable with an MTF because then the reflected or equivalent I-parameter depends on the control signal.

Then on slipping the part inductance (I22) depends on the transformation ratio, that is on the control signal α. Other remedies are then preferable.

The problem of Fig. 6.16 occurs frequently in robotics (sec. 8.1) and mechanics in general, if 2 I-elements are separated by a modulated transformer that expresses the variable speed ratio of linkages. This ratio depends on the angle of some joints.

6.5.1 Additional C- and I-elements or the Remedy of the Bad Conscience

An efficient way to break derivative causality is to add a small I-element I22 between the C-elements in Fig. 6.17, top. Alternatively a small C-element C22 can be added between the I-elements as in Fig. 6.17 below. This corresponds to a compliance (spring effect) in the arms of a robot or other linkage.

The location of the added C-element at the right or left of the transformer determines the causality. In the case of modulated transformers it is often important because the transfer equations may be much easier to write and to program in one causality than the other. It is a part of the art of robotics.

No bad conscience should be required as long as the added elements correspond to real physical effects. However, if they are small, they carry very small parameters which lead to prohibitively small time steps in some cases. The remedy of the bad conscience is to arbitrarily increase the parameters for an acceptable time step.

The effect of inflating (increasing) the small parameters means that one really simulates a different system from the one of interest. The author sometimes had an uneasy feeling or bad conscience and asked mathematicians about the procedure. No practicable answer was obtained except to try it out. A good trial procedure is to run up the simulation with the inflated parameters. When it looks good, make a part time simulation with original parameters and small time steps, perhaps over lunch.... If the results look the same, there will probably be no trouble.

An excuse for the inflated parameters is that the original ones produce very high frequency oscillations or EV, which are hopefully well damped. If not, they might still not be representative, because other high frequency effects might be neglected on the Bondgraph. It is around here where the usefulness of the

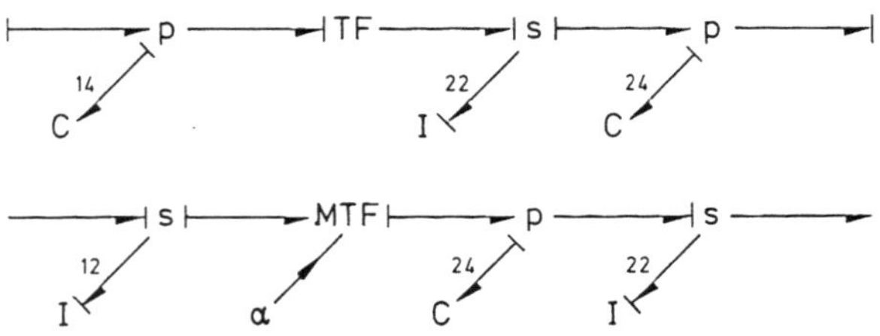

Figure 6.17 Remedy to derivative causality by adding small I22 (top) and C24 (bottom). Doubtful are the parameter values, which force short time steps when small.

simulation reaches its limits and testing of a real machine or prototype comes into focus.

There is some parallel with the steady state Bondgraph of Fig. 6.5. If we are interested in steady state and medium frequency response, especially for control purposes, we can neglect or deform the high frequency behaviour by inflated parameters. The final decision and interpretation depends on experience and the intuition of the human commander.

6.5.2 Calculation with Derivative Causality

We can also "take the bull by the horns" and calculate within derivative causality. In TUTSIM this requires a block DC (derivative capacity) or DI (derivative inductance). There is one important limitation inherent in derivative algorithms. The element in derivative causality, f.i. the mass, must be smaller then the element remaining in integral causality. This is a convergence condition.

Figure 6.18 explains the derivative causality of I14, left as Bondgraph and right as blockdiagram. As visible on the Bondgraph, its loop gain is I14/I24 and this must be smaller than one. Otherwise the calculation diverges. The situation is similar to algebraic loops which we shall treat right after the robot example.

If the I-elements are separated by transformers or modulated transformers, one element must be slipped over before stability condition applies. In the case of an MTF, the stability condition may be violated at certain values of the signal and computation impossible from there.

6.5.3 Example: Hydraulic Robot

Figure 6.19 shows a hydraulic robot as an example of 2 masses connected by a modulated transformer. There is a mechanical movement with a variable ratio linkage between $\dot{x}11$ (piston speed) and $\dot{x}22$ (vertical load speed). The Bondgraph has no linkage compliance, so I11 is chosen to be in derivative causality. Since it represents a piston mass much smaller than the load mass I22 the convergence condition (derivative mass smaller) is fulfilled over the range of interest.

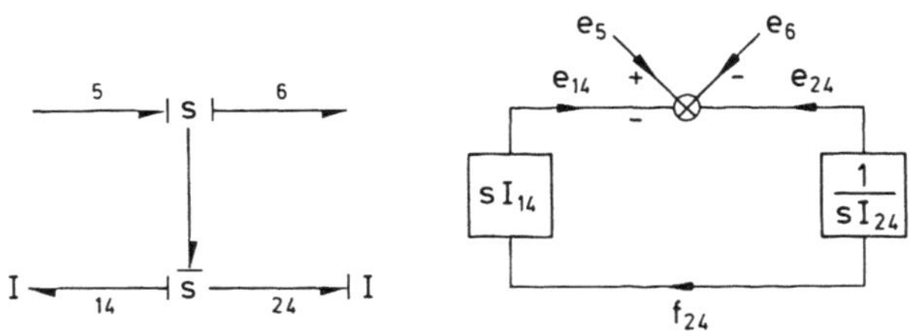

Figure 6.18 Part Bondgraph and blockdiagram explaining the loop gain condition with derivative I-elements. The derivative mass must be smaller than the (reflected) integral mass.

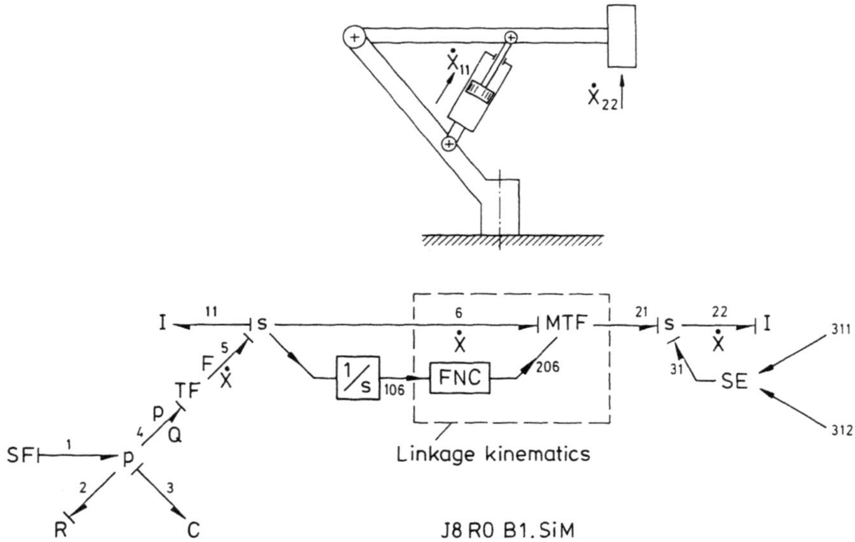

Figure 6.19 Schematic and Bondgraph of a hydraulic robot with variable linkage ratio in MTF. It is generated through the function generator FNC206 from the position x106.

Figure 6.20 shows the start up, with a shock load of 10 kN applied after 0.05 secs. and removed after 0.4 secs. Both load shocks introduce oscillations, the removal more violent and less damped because of the different linkage ratio.

Figure 6.21 shows a parameter check-list and Fig. 6.22 the TUTSIM listing of our robot simulation; both should be compared. The use of such parachecks is explained in sec. 6.6.

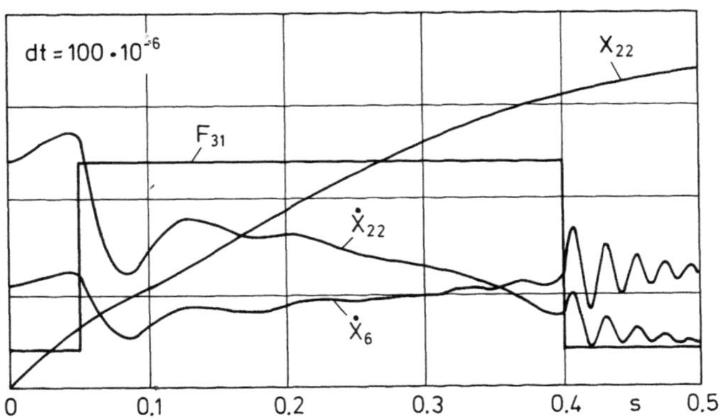

Figure 6.20 Simulation of the hydraulic robot with a shock load applied after 0.05 sec. and removed after 0.4 sec. Both introduce transients, but with load removal stronger, faster and less damped ones. This is a consequence of the different linkage ratio.

Robot paracheck (file J8ROB1)

Element		Valeur	Unit	Parametre output	valcode
SF_1	servovalve	800	μLA		
R_2	parallel resistance	0.25	$\frac{MPa}{\mu LA}$	Q_2	4.0
C_3	hydraulic compliance s = 1m	71.5 E-03	$\frac{cm^3}{MPa}$	p_3	71.5 E-03
$TF_{4,5}$	piston area	400	mm^2	F_5, Q_4	400
I_{22}	load mass	100	$\frac{Ns^2}{m}$	$\dot{X}_{22}$	
I_{11}	parasitic mass	5	$\frac{Ns^2}{m}$	F_{11}	5.0
MTF_6	linkage x_6 = 0.1 0.2	0.25 0.3	m	0.35 0.4	0.45 0.50
FNC_{206}	ratio r = 0.45 0.5	0.6 0.7	m	0.8 1	1.4 2

Figure 6.21 Paracheck of the hydraulic robot. Here the compact hydraulic units of sec. 6.6 have been used.

Apart from the intrinsic interest this example will help the reader to experiment with derivative causality in TUTSIM.

6.5.4 Algebraic Loops

Algebraic loops are an important difficulty which should be avoided in TUTSIM. Large computers can handle them by a time consuming subroutine (Schmidt 1980, sec. 4.4.1). In fact, there is a causal loop without integrators, or a closed loop in the corresponding blockdiagram. We have the advantage of easily recognizing such troublesome loops on a Bondgraph.

Figure 6.23 shows at the top a very simple Bondgraph, yet it has an algebraic loop between R2 and R4. To recognize it we either:

1. Follow the needles (fine points of the bond) out from R4 as flow travels over both junctions to hit R2. This builds a causal loop without integrators.
2. Follow the pistons (the end of the bond with the causal stroke) out from R2 as effort travels over the junctions into R4: This also indicates an algebraic loop.

Needles or pistons ending on integrators build a non algebraic loop and are harmless: they build integrating loops. Further, needles or pistons ending on sources do not build loops at all, since sources give no return signal.

The loops become more clear in the partial blockdiagram of Fig. 6.23 below. The signal has to travel over both R2 and R4 blocks without meeting an integrator in between: that constitutes an algebraic loop.

```
Timing: 100.000E-06  ,DELTA  ;   0.5000000   ,RANGE
PlotBlocks and Scales:
Format:
     BlockNo,  Plot-MINimum,   Plot-MAXimum;  Comment
Horz:     0 ,   0.0000     ,    0.5000000  ; Time
  Y1:    22 ,   0.0000     ,    5.0000     ; load mass
  Y2:   122 ,   0.0000     ,    1.0000     ;
  Y3:     6 ,   0.0000     ,    5.0000     ; linkage speed ratio
  Y4:   313 ,  -2.000E+03  ,   18.000E+03  ; sign invert

 600.0000            1 SF      101                    ;hydraul flow source
   4.0000            2 R         3                    ;leakage resistor
   0.0715000         3 C         1      -2       -4   ;hydraul capacity
   0.0000
 400.0000            4 TF        6                    ;piston arrea
 400.0000            5 TF        3                    ;piston force
                     6 MTF     206      22            ;linkage speed ratio
   5.0000           11 DI        6                    ;piston mass
   1.3500
   0.0000
                    21 MTF     206       5      -11   ;linkage force ratio
 100.0000           22 I        21      31            ;load mass
   3.0000
  -1.0000           31 SE      311     312
   0.0000           32 CON
   1.0000          101 CON                            ;file bonsi65
   0.0000          106 INT       6                    ;mech integrator
   0.0000          122 INT      22
                   206 FNC     106
     1     0.0000             0.4500000
     2     0.1000000          0.4500000
     3     0.2000000          0.5000000
     4     0.2500000          0.6000000
     5     0.3000000          0.7000000
     6     0.3500000          0.8000000
     7     0.4000000          1.0000
     8     0.4500000          1.4000
     9     0.5000000          2.0000
    10     1.5000             8.0000
  0.0500000        311 PLS
  1.0000
 10.000E+03
  0.4000000        312 PLS
  1.0000
-10.000E+03
 -1.0000           313 GAI      31                    ;sign invert
```

Figure 6.22 TUTSIM listing of the hydraulic robot. Comparison with the paracheck will eliminate most parameter errors.

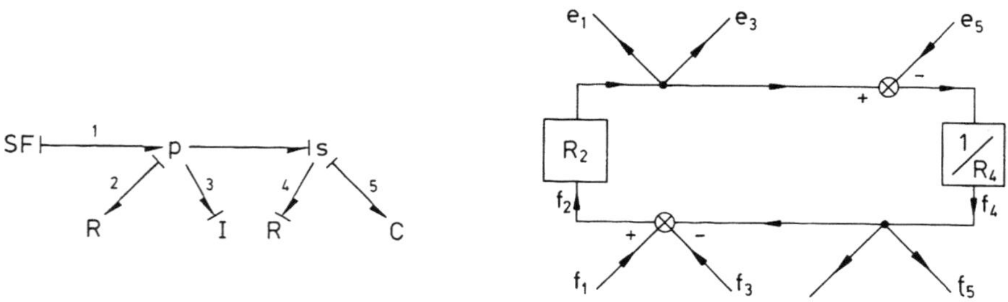

Figure 6.23 Simple Bondgraph with an algebraic loop between R2 and R4. Below a blockdiagram displaying the loop with static gains.

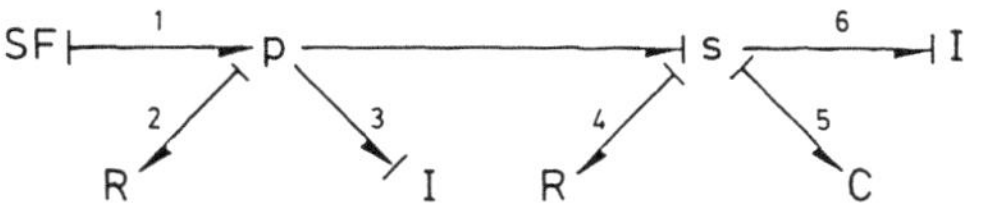

Figure 6.24 Modification of Fig. 6.23 by the small added inductance I6; it drives R4 into resistance causality and effectively breaks the algebraic loop.

Figure 6.24 shows a remedy: Simply by adding I6, the R-element R4 goes from conductance into resistance causality and no more algebraic loop exists. The flow hitting R2 comes now out of I6, an integrator, so all is well.

The question is again the physical basis of I6 and its parameter value. If it is very small, we might be tempted to inflate it (make it larger than physically justified) and we are back to the bad conscience discussed with derivative causality above.

In TUTSIM an algebraic loop can be calculated by the ADL block (Algebraic DeLay), if the loop gain is smaller than 1. No certain information exists about the effect or possible loss of accuracy of this procedure.

The nice thing is that algebraic loops can usually be eliminated when writing the Bondgraph by ingenious changes when it tends to develop. Such loops are certainly a major nuisance, probably more then derivative causalities.

6.6 Compact Parameters and Paracheck

Digital computers are said not to require scaling of variables as do analog computers. This is only partly true and especially in hydraulics we need a limited scaling with compact parameters or units. Otherwise parameters of the order of 10^{12} will lead to inaccuracies or overflow/underflow errors, even in big machines. In hydraulics, the pressure unit Pa is too small and the flow unit, m^3/sec is impractically large. So we use MPa and cm^3/sec, that is we change the basic units by a factor of one million. In this way power remains invariant since 1 W remains 1 MPa times 1 cm^3/sec. The same is true of time constants, resonance, frequencies and EV (Eigenvalues).

For a computer it makes quite a difference, if a pressure is represented say by 12 (MPa) or by $12*10^6$ (Pa) similarly with flow, since the computer takes no account of the units. Still more impressive are the one-ports, resistances capacitances and inertias, which change their number by a factor of 10^{12}. As an example for compact units, we go from $2*10^{-12}$ (m^3/Pa) to 2 (cm^3/M Pa) with a typical hydraulic capacitance.

Figure 6.25 contains a table of the compact units for the one-ports and the transformer with the compact pressure and flow units. Let us mention that for a ram the effective area has to be given in mm^2, and for hydrostatic machines the displacement volume in cm^3/rad. Both are the ratios or parameters of transformers in appropriate causality.

si units	R	C	I	TF-lin	TF-rot
pure	$\frac{Pa\,s}{m^3}$	$\frac{m^3}{Pa}$	$\frac{Pa\,s^2}{m^3}$	m^2	m^3/Rd
compact	$\frac{MPa\,s}{cm^3}$	$\frac{cm^3}{MPa}$	$\frac{MPa\,s^2}{cm^3}$	mm^2	cm^3/Rd

Figure 6.25 Table of compact hydraulic units. Pressure and flow units are changed by a factor of one million, but times, EV and resonance frequencies remain the same. The parameters of the oneports have to be changed by a factor of 10^{12} giving reasonable values for hydraulic capacitances and resistances.

6.6.1 Compact Mechanical Units

With heavy machinery it is sometimes useful to also employ compact mechanical units in addition to the compact hydraulic units, which are almost always essential. The best choice is to express force in heN (100 N) and speeds in cm/sec (0.01 m/sec). Hence, frequencies, time constants and Eigenvalues again remain unchanged. Rotary mechanical units are never changed, and remain thus at Nm and rad/sec, which we like to call Pz (Pulz).

As the table in Fig. 6.26 shows, here the one-ports have to be changed by a factor of 10^4 that is to heN*sec/cm for resistances, cm/heN for capacitances (springs) and heN*sec^2/cm for masses (corresponding to 10^4 kg). Further the wheel, a linear/rotary transformer needs a radius in cm. The ram area with both hydraulic and mechanical compact units is in cm^2.

	R	C	I	SE	SF	Cylinder TF-lin	Pump/motor TF-rot
Hydraulic components	$\frac{MPaS}{cm^3}$	$\frac{cm^3}{MPa}$	$\frac{MPaS^2}{cm^3}$	MPa	cm^3/s	cm^2	cm^3/Rd
	$\frac{MPa}{\mu LA}$	pLF	TLHe				
	R	C	I	SE	SF	Wheel TF lin-rot	
Mechanical components	$\frac{heNS}{cm}$	$\frac{cm}{heN}$	$\frac{heNS^2}{cm}$	heN	$\frac{cm}{s}$	cm	
			10^4 kg				

Figure 6.26 Table for compact mechanical units to be used together with compact hydraulic units. Here force and speed are changed by a factor of one hundred, again leaving Eigenvalues constant. Wheel radius has to be given in cm and ram areas in cm^2 (also when compact hydraulic units are used).

In mechanics there are not enough names for the many units, which are so abundant in electronics. Therefore, there is the possibility to use electric names and abbreviations preceded by an L for liquids. Hence we can call normal, non compact units:

LA = liquid Ampere = 1 m^3/sec
LF = liquid Farad = 1 m^3/Pa
LOh = liquid Ohm = Pa/(m^3/sec)

Liquid volt is not needed since we have the name Pascal for it. Introducing now compact hydraulic units, pressure is expressed in MPa (Megapascal), flow in μLA (micro LA), capacitance in pLF (pico LF) and resistances in TeLOh (tera LOh).

Once familiar, this is an effective terminology especially for people with experience in electronics.

6.6.2 Paracheck

Writing a Bondgraph is usually easy, finding the parameter values is difficult. In fact, parameter values require 80% of the effort and produce 90% of the errors. It is therefore good to use a standard parameter checklist called briefly paracheck. For interpretation one checks along the chain:

Schematic-Bondgraph-paracheck-TUTSIM listing-result.
Thus parameter errors are much easier to detect.

For the construction of the paracheck refer to the example in Fig. 6.21. There are 2 columns at the right, since the parameters for resistance and transformers depend on causality. One column contains the output variable and the other the value code to be entered. Note especially that for an R-element in conductance causality we have to enter the reciprocal resistance or conductance. Transformers are another element for which the parameters for a given piece of hardware depend on causality. If the causality of a transformer is reversed, the parameters become reciprocal.

6.6.3 Practical Run Up and the Bondgraph Tree

It is dangerous to start with a complex simulation even if supported by a Bondgraph and it is also inefficient with working time. We prefer to start with a

Figure 6.27 Bondgraph tree showing the systematic building of complex simulations from simple ones of the same physical system. The results support each other as would the branches of a tree.

simple Bondgraph, run it and check the results carefully for correctness. When acceptable, we pass to the next level by adding elements and examine how the results change. Thus, the simpler simulation levels support the more complex ones, just as the branches of a tree support each other. We speak then of a Bondgraph tree (Fig. 6.27), which is our building method for complex simulations.

It should be noted that the term "tree" denotes an entirely different concept in linear graph theory. Here it simply means several simulations of the same fundamental problem in order to check the results.

Obviously we do not write the complex Bondgraph anew, but obtain it by editing (modifying) the preceding one. This is a quick and pleasant procedure on a computer.

As mentioned, a Bondgraph should be build in stages and a further precaution is to give a value of one to all parameters at first, except resistances and conductances which receive a value of 0.2. From our formulae in sec. 6.3 the resonance frequency will be 1 rad/sec. and the damping will be 0.1. The maximum time step would be 0.25 but we normally use a smaller one, say 0.1. This shows that the Bondgraph is correctly wired up before the real parameter values are entered. Naturally, real parameters may require undesirably small time steps and require the remedies discussed above, but this is the practical life of a simulation expert.

6.6.4 The Art of Labelling

As we have seen on our examples, the reference numbers or labels designate essentially the bonds or connections. The elements are designated by implication: they receive the label of their bond, as f.i. C23. Elements in signal connections and in blockdiagrams receive the label of their out-going connection (f.i. GAI41). Two-ports are referred to by two labels, as f.i. TF29, 31 (all on Fig. 4.16). Different labelling schemes are possible and used by other authors.

Reference numbers in TUTSIM require one more point, since the program cannot distinguish between efforts and flows. So the bond label designates the variable, effort or flow, that travels from the element to the next junction according to the causality stroke (compare Figs. 2.24 and 2.25). This convention would run into trouble if two elements were connected without junctions, but then we interpose a dummy junction with two bonds, which is an identity.

In principle the labels of the bonds are arbitrary and can run from 1 to 9999 in TUTSIM, but alphanumeric labels, such as 23a are not admitted, unfortunately. In practice a good labeling scheme contributes much to the ease of interpretation of a simulation; we have endeavored to use good and easily remembered labels in our examples.

As a guiding rule subgroups of nearby elements should carry similar labels. Thus we assign a dekade of labels with 2 digits to a subgroup, say 11 to 19, 21 to 29 and so on until 91 to 99. A third digit is added to give a detail, especially with bloc-bondgraphs, as f.i. 231 helps to build F23 on Fig. 8.14. Labels with 4 digits are avoided if possible since they are hard to remember. It follows that under no circumstances should bonds be labelled consecutively, as done in more mathematically oriented publications. If we had to add an element to a subgroup it would get the next free label, which is usually very different: an obstacle to remembering and interpretation.

As a refinement, the second digit indicates the kind of element; on Fig 91-1b we find C13 and C23, R25 and R35 and I24. Further C-, R- and I-elements would receive second digits 3, 5 and 4, respectively. The more repetitive a Bondgraph, the more useful such a scheme becomes. Another method is to give to elements between two subgroups the second digits of 9 and 1, that is for the last and the first of the next subgroup, respectively. An example is Fig. 4.16 (hydrostatic transmission) with MTF19, 21 between input mechanical and hydraulic elements and TF29, 31 between hydraulic and output mechanical elements. Finally, the time bloc TIM receives 999, the highest possible label of older TUTSIM programs.

Nevertheless, the larger the Bondgraph and the more complex the system, the more difficult a good labelling scheme becomes. Definitely, thinking about labels at the beginning of a simulation is time well spent; changing labels later is almost impossible.

6.7 Causal Loops and Inspectional Analysis

As we have seen, the system designer should not rely blindly on EV (Eigenvalue) programs, but determine and check them independently by inspection. The method of causal loops gives a good approximation to EV, hence resonance frequencies, time constants and time steps; it is an alternative to the method based on electric circuits described in sec. 6.4.

A causal loops comes from the idea that the signals, effort and flow travel on either side of the bonds, as we have indicated in secs. 2.3 and 2.5. We can visualize this better by writing small (full) arrows on the side of the bonds (compare Fig. 2.13); strictly they are redundant since they give the same information as the causal strokes.

Figure 6.28 contains both the side arrows and the causal strokes and will illustrate the concept of causal path and causal loops. The example could be part of a hydrostatic transmission similar to Fig. 4.16 with an added Lanchester damper build from R34 and I38 (compare Fig. 3.11); however it is general and can refer to any type of effort and flow. A causal path is then a sequence of side arrows from an origin, usually from a one-port towards a destination (end point), usually another one-port, passing over some intermediate elements. We have the following detail points:

- A causal path is direct if it passes only over junctions.
- A causal path is indirect if it passes also over transformers (TF) or gyrators (GY) as f.i. over TF29, 31 between C27 and I32.

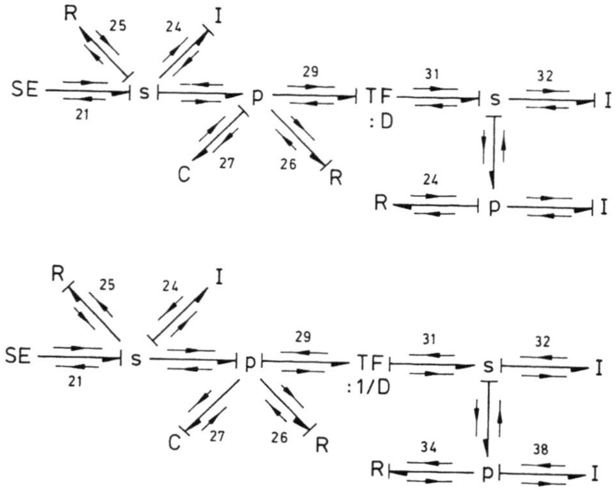

Figure. 6.28 Hydrostatic transmission (part with Lanchester damper in integral (top) and derivative (bottom) causalities. The small side arrows indicate the direction of action of the signals. These arrows carry the same information as the causal strokes but help to visualize the the causal path, which are a sequence of side arrows. The EV can be seen immediately from the causal loops.

- A causal path is extended if it goes also over one-ports. Thus the variables pressure p29 and flow Q29 are connected by an extended causal path over TF29, 31 and I32.

A causal loop is build from 2 causal path, direct or indirect, that build a closed loop, going from one element, usually a one-port to another element and back. Examples are between I24 and R25 or between C27 and I32. Transforming the Bondgraph into a blockdiagram, causal loops become closed loops in the sense of control theory; they determine the dynamical behaviour of the system.

Since causal loops of a Bondgraph indicate the loops of the corresponding blockdiagram, they relate to the modes of the system and each loop determines approximately an EV. For the analysis by inspection we identify the causal loops and associate to each a time constant or a resonance frequency, depending on whether it runs between C- or I- and R-elements or between C- and I-elements, respectively. Thus we find easily on Fig. 6.28 the Eigenvalues 1/(R26C27), R25/I24 and D^2/(C25I32). The required time step follows imediately (sec. 6.2).

Naturally, the frequencies and time constants will be somewhat detuned by coupling with the other modes of the system, but this is usually small, especially with high, troublesome EV.

More formally, the approximate EV comes from solving an equation of the following type for each causal loop:

$$1 + \text{loop gain} = 0.$$

This is related to the Mason's rule for the transfer function of closed loop systems.

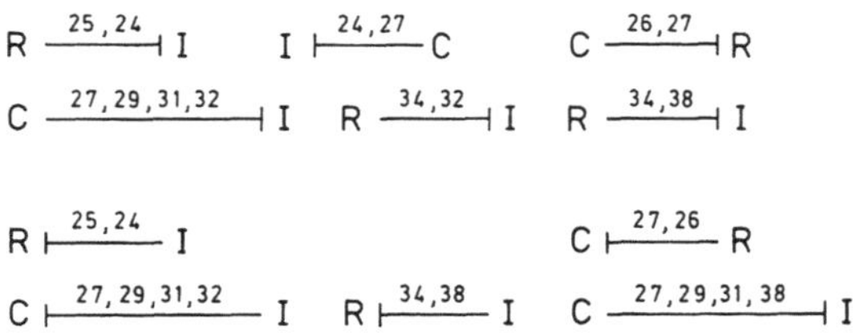

Figure 6.29 Concise representation of the causal loops of Fig. 6.28 for integral (top) and derivative (bottom) causalities. Some loops remain the same on change of causality and give an especially good estimation of EV.

The lower part of Fig. 6.29 shows the same Bondgraph with derivative causalities on all energy storers. We see that some causal loops remain the same, except for reciprocal loop gains, but some other disappear or appear. In order to show more clearly the change, Fig. 6.29 represents the causal loops obtained from integral causality on top and from derivative causality on bottom. Based on computation experience it appears that the loops that remain the same after a change of causalities give an especially good estimation of the full EV. In other words, loops invariant under causality change are strongly coupled to the full EV.

The computational problems of sec. 6.5 can also be identified from the causal loops:

- A causal loop between two C- or I-elements leads to derivative causality (if no gyrator is run trough).
- A causal loop between 2 R-elements leads to an algebraic loop.
- If the causal loop includes a gyrator, derivative causality will appear if it runs from a C-element to a I-element (more generally: if the loop includes an odd number of gyrators).

The signs of the causal loop gains follow from the power orientations at the junctions: through-going orientations give no sign of change, but two inward or two outward orientations do so.

6.7.1 Controllability and Observability

The concepts of controllability and observability were introduced in the 1960's by Kalman (Takahashi 1970); they refer to the state variables of complex control systems. They can be seen on the Bondgraph (Sueur-Dauphin 1990) as follows:

- A state variable is controllable if there is a causal path between input and its C- or I-element; in Fig. 6.28 it could be between the effort source and I24 for the momentum p24.
- A state variable is observable if there is a causal path between its C- or I-element and the observer, in Fig. 6.28 between I32 and the signal connection (full arrow) at right.

A causal path represents essentially the forward gain in Mason's formula. Therefore the full gain between state variable and output (or input for controllability) is difficult to determine, since it requires division by the loop gains according to Mason's formula for multi-loop systems. This formula, an extension of the simple one-loop formula is involved to formulate, but seems easy to apply. What remains is that the system is inobservable (incontrollable) if there is no causal path between the state variable and the observer (between input and state variable, respectively).

For the practical engineer, the degree of observability or controllability can be estimated by the length of the causal path:a short causal path makes it easy but a long, indirect or even extended path makes observation and control difficult, perhaps even impractical. Returning to general computational problems, a recent thesis gives a good view of integration routines, time steps and a new simulation programm from Bondgraph models: J.F. Broenink, Computer-aided Physical Systems Modelling and Simulation: a Bondgraph Approach, Department of Electrical Engineering, 1990, University of Twente, NL-7500 AE Enschede.

7 Applications to Thermodynamics, Chemistry and Biology

7.1 Meet and Love Entropy

Entropy often appears as a mysterious concept yet it is really quite simple: Entropy flow is the thermal flow variable in the sense of Bondgraphs (Table 2.1) and entropy itself is accumulated entropy flow (Thoma 1970B). Hence the fundamental point to know is that thermal power in conduction, usually called heat flux, is absolute temperature times entropy flow. This fundamental equation appears in Fig. 7.1. Some important observations:

- The fundamental equation is strictly valid for conduction of thermal power (heat flux) without mass flow.
- With mass or convection there is a correction to the fundamental equation, which remains approximately valid. It is expressed by the convection factor of sec. 7.2.
- We have to take the absolute temperature (expressed in Kelvins) in the fundamental equation.
- Entropy is often associated with disorder. This is true but not necessary for most practical engineering including Bondgraphs.

More in detail about disorder, this is a detailed structural theory of entropy, needed in physics; engineering works well with the above "phenomenological" notions.

If we compare electrical engineering, nobody really knows what electricity is, or in other words, there is no structural theory of electric charge, the fundamental concept. Continuing this line of thought, entropy is thermal charge in analogy to electric charge. Yet everybody can work in comfort with electricity, because he knows the essential properties of electric current (electric charge flow). So we need to know the essential properties of entropy (thermal charge).

7.1.1 Properties of Entropy

- Entropy is a kind of packing material or transport substrate for thermal power, just as electric charge. Without it there can be no power transport.
- Entropy, just like electric charge, is conserved in all friction free processes, that is those in C, TF,..., elements and junctions, etc.

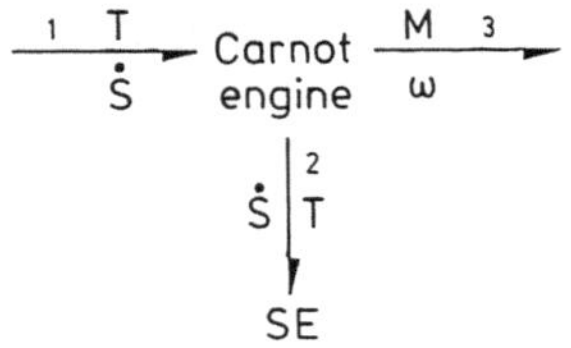

Figure 7.1 Fundamental equation of thermal conduction, where power or heat flux equals entropy flow times absolute temperature.

– Only in friction processes, unlike electric charge, is entropy generated. Friction takes place only in R-elements.

In the thermodynamic literature processes without friction are called reversible or thermodynamics of equilibrium. With friction, processes are called irreversible or thermodynamics of non-equilibrium. This fits well with our statement in sec. 2.2 that all Bondgraph elements except R-elements are reversible.

7.1.2 Fundamental Non-friction Thermodynamics

As an example of the conservation of entropy in non friction processes, Fig. 7.2 shows an ideal engine for conversion of thermal into mechanical power. Thermal power (heat flux) arrives at high temperature T1 and should be converted into rotary mechanical power in bond 3. Due to entropy conservation, the same entropy flow must be removed from the engine at the lower temperature T2 and requires a corresponding power. Only the difference of both thermal powers is available for conversion into mechanical power.

Since there is no friction in Fig. 7.2, by supposition, the entering and leaving entropy flows are equal. Dividing output mechanical power by input thermal power an expression for the efficiency of this ideal engine appears on the last line of Fig. 7.2. This is the celebrated Carnot efficiency formula contained in his book "On the Motive Power of Fire", Paris, France, 1824. The Carnot engine is perfectly reversible in the sense that it can pump entropy from lower to higher temperatures as done in domestic refrigerators.

7.1.3 Frictionless Thermodynamics in C-Fields

Continuing with frictionless thermodynamics, Fig. 7.3 shows a piece of matter, perhaps an ideal gas as a 2-port C-field with 2 bonds:

$$\dot{Q}_1 = T_1 \dot{S}_1 \qquad \dot{E}_3 = \dot{Q}_1 - \dot{Q}_2$$

$$\dot{Q}_2 = T_2 \dot{S}_2$$

$$\dot{S}_2 = \dot{S}_1 \qquad \frac{\dot{E}_3}{\dot{E}_1} = \frac{T_1 \dot{S} - T_2 \dot{S}}{T_1 S} = \frac{T_1 - T_2}{T_1}$$

No friction

Figure 7.2 Ideal engine, as described by Carnot, for conversion of thermal into mechanical power. Since entropy is indestructible, it has to be removed using some power to push it into the lower source. Then the efficiency formula follows. Non-ideal engines would have some friction and require more power for its removal and have lower efficiency.

Figure 7.3 Matter as 2-port C-field with pressure, volume, temperature and entropy as variables. In case of variable mass, a third bond has to be added.

– Thermal bond with $\dot{S}$ and T as effort and flow.
– Hydraulic bond with p and $\dot{V}$ as effort and flow.

The causalities of Fig. 7.3 will jump into the eyes of a Bondgrapher: The mandatory causalities of the effort sources impose derivative causalities on the C-field. This is true and acceptable for such a conceptual Bondgraph. In order to make it computable, some of the remedies of sec. 6.4 need to be applied.

Further bonds can be applied, in particular magnetomotoric force and magnetic displacement useful in supraconductivity. Further, there is chemical potential and mass flow as we shall use in secs. 7.2 and 7.3. It must only be assured that friction effects are neglegible.

The essential idea of the C-field in Fig. 7.3 was contained in the book by Carnot. So he already knew about entropy, which he called caloric in distinction to heat (french chaleur), but the name entropy was introduced later, about 1850 by R. Clausius.

The 2-port C-field has 2 degrees of freedom. This means that the external part of the system can apply 2 variables, f.i. pressure and temperature, but then the C-field determines how much entropy it will absorb and what volume it will take up. With variable mass a third bond and degree of freedom appears having the chemical potential as the effort variable (secs. 7.2 and 7.3).

As always, the 2-port C-field of Fig. 7.3 is energy conserving and obeys the Maxwell relations. The natural input variables are entropy and volume, which lead to integral causality, whence the C-field calculates pressure and temperature. Sometimes, both for human thought and in textbooks, different input variables are preferred, say pressure and temperature. In the constitutive equation, entropy, that is stored entropy, depends primarily on temperature but also on pressure due to the cross connections of C-fields. Generally, by increasing pressure at constant temperature stored entropy decreases, it is "squeezed out by the pressure".

The relations in a thermal C-field are similar to those of the capacitor with moving plates of Figs. 5.8 and 5.9. The celebrated Carnot cycle is the same as on Fig. 5.9, only with temperature and entropy flow instead of voltage and electric current. Here the 2 phases in connection with the temperature source are called isothermals, and intermediate 2 phases, where the thermal bond disconnected, are called adiabatics. They avoid thermal short circuits as vividly described by Carnot. The switches of Fig. 5.9 can still be recognized in the valve gear of modern automobile engines.

So for a Bondgrapher, entropy is a very clear concept that he can love, because it helps to understand nature. Heat, on the other hand, is more involved, because it is not conserved over a cycle in a C-field. So why is heat used so frequently in thermal engineering and heat exchanger design? We shall see this from the thermodynamics of frictions.

7.1.4 Thermodynamics of Friction Elements

All irreversibility takes place in frictions or R-elements. They are power conserving in the sense that the dissipated power, absorbed effort times flow, appears as heat flux, or more precisely as newly generated entropy flow times absolute temperature. We take here friction in a wide sense as follows:

1. All normal frictions in mechanical, electric and hydraulic resistances.
2. Entropy (heat) conduction at substantial temperature drop.
3. Diffusion in physics and chemistry.
4. Chemical reaction resistance (sec. 7.3)

Disposing quickly of the last two types, diffusion plays no further role in this book and chemical friction will appear briefly in sec. 7.3. Speaking now about type 1 friction, it takes place in all R-elements. They are power conserving and a source of entropy flow. Consequently, all R-elements are really 2-port RS-fields as shown in Fig. 7.4. A preferred causality is indicated, but in the non-thermal bond either causality can be used. Frequently there is a cross coupling in an RS-field, f.i. when the resistance depends on temperature.

Figure 7.4 also contains the equation which follow from power conservation. The generated entropy flow $\dot{S}$ equals voltage times current divided by absolute temperature (or in general non-thermal effort times flow). As stated before, it is essential to take the absolute temperature, thus the Kelvin scale and not the Celsius scale. This is one particular point of thermodynamics.

Figure 7.5 refers to thermal or entropy conduction, number 2 above. This is power conserving and input heat flux equals output heat flux. So entropy flow increases as the temperature drops. At the top we have a schematic, say a part of a wall of a heat exchanger with 2 control areas having or passing the indicated temperatures and entropy flows. The RS-field, top right, also represents this thermal conduction.

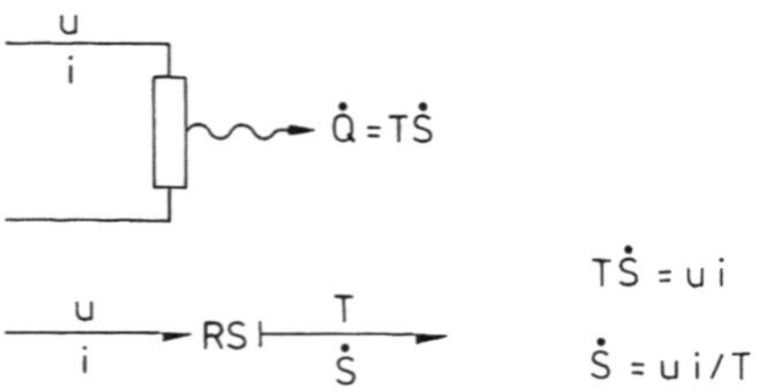

Figure 7.4 An R-element, like an electric resistor, is power conserving and generates entropy flow.

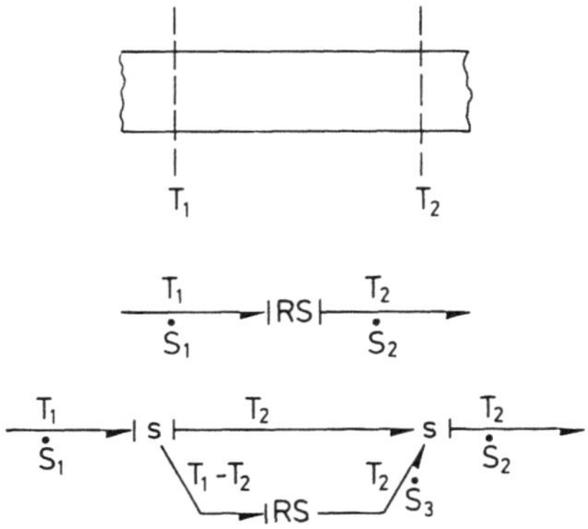

$$\dot{S}_1 = \frac{K_1}{T_2}\,(T_1 - T_2)$$

$$\dot{Q}_1 = K_1\,(T_1 - T_2)$$

$$\dot{S}_3 = \dot{S}_1 (T_1 - T_2)/T_2$$

Figure 7.5 Cross-section through a wall with thermal conduction with 2 control areas (top). Center: Reticulation as RS-field driven by the temperature drop. The new entropy flow is injected to the outflow.

The lower part of Fig. 7.5 contains a detailed reticulation of conduction and entropy generation. According to the selected causalities, the temperature differences drives the entropy flow in the RS-field. We have chosen in the equations the constant of proportionality for entropy conduction as K1/T1, so the equation for heat flow is K1 times temperature difference, as usual in thermal conduction. K1 is then the power conductivity.

Comparing entropy conductivity and power conductivity, they are related by the equations in Fig. 7.5. Both conductivities depend on temperatures, since K1 depends on temperature in general. It is not that power conduction is more natural, but it is more convenient for the following reason.

In conduction entropy flow is not constant (divergence free in the sense of vector calculus), but power flux is. Therefore, the known partial differential equation between heat flux and temperature is simple and suitable to calculate temperature distributions. There exists a partial differential equation between entropy flow and temperature but it is impractically complex.

For a calculation with TUTSIM, Fig. 7.6 shows a suitable Bondgraph derived from Fig. 7.5. We have here a simple electric (or generally non-thermal) resistor and a separate flow source. Its output is multiplied with the voltage and divided by T2 to determine the output entropy flow $\dot{S}2$. A practical application in

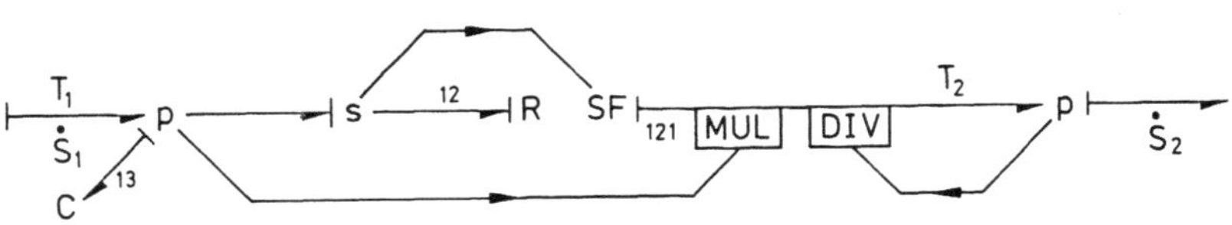

Figure 7.6 Reticulation of entropy conduction for TUTSIM. The equations of the preceding figure are implemented by the MUL and DIV blocks.

connection with possible overheating of hydropneumatic accumulators appears with the worked examples of sec. 9.2. The Bondgraphs of Figs. 7.5 and 7.6 are only true if the power direction does not change. Otherwise they must be completed to indicate the fact that the new entropy is always injected at the lower temperature side, see Fig. 9.4b.

7.1.5 Pseudo-Bondgraphs for Heat Flux

Since the reticulation of entropy flow is complex, even if intellectually satisfying, one can use Pseudo-Bondgraphs for problems with heat conduction. Here flow is taken as heat and power flux and effort as temperature. We then have capacitors with heat energy as a function of temperature and simple power conducting R-elements. The following example illustrates their use.

7.1.6 Simulation of a Silicon Slab

Figure 7.7 shows a cross-section through a silicon slab and an electric circuit, below the Pseudo-Bondgraph. Here heat power flux is taken away as flow through RS14 multiplied by voltage u12. The following C and R network models heat capacity and power conduction of the material.

Figure 7.8 shows the corresponding TUTSIM listing and simulation. Using Pseudo-Bondgraphs of this kind is useful in practice but definitely not beautiful.

7.2 Convection, Matter Flow and Solar Energy

Transport of energy by flowing matter or convection was excluded from sec. 7.1. It is a difficult subject related to Eulerian and Lagrangian frames of reference in

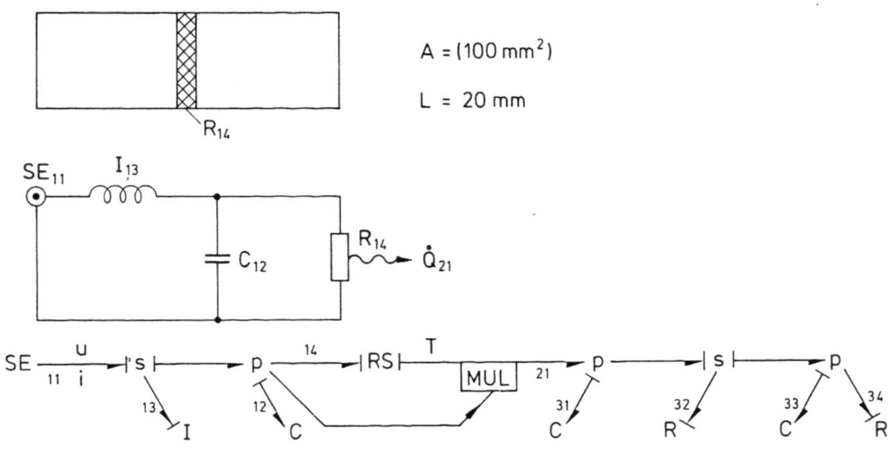

Figure 7.7 Cross-section, circuit and Pseudo-Bondgraph for heat conduction in a slab of silicon.

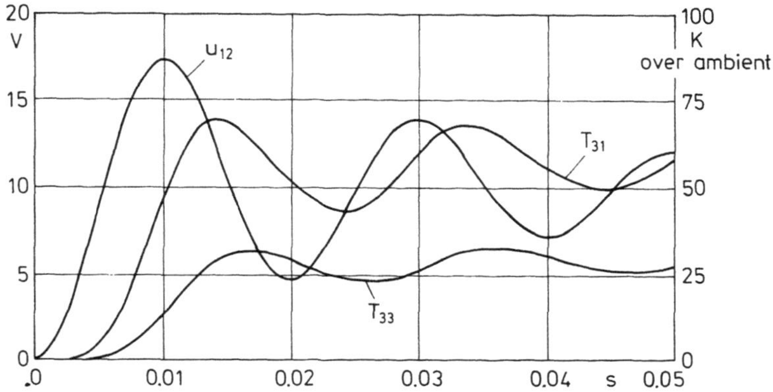


```
Timing: 100.000E-06   ,DELTA  ;    0.0510000   ,RANGE
PlotBlocks and Scales:
Format:
     BlockNo,   Plot-MINimum,   Plot-MAXimum;   Comment
Horz:      0 ,   0.0000       ,    0.0500000  ; Time
  Y1:     12 ,   0.0000       ,  20.0000      ; electric capacity
  Y2:     31 ,   0.0000       , 100.0000      ; thermal capac #1
  Y3:     33 ,   0.0000       , 100.0000      ; thermal capac #2
  Y4:        ,                ,               ;

  10.0000            11 SE       111                   ;voltage source
   0.0010000         12 C         13      -14          ;electric capacity
   0.0000
   0.0100000         13 I         11      -12          ;inductance
   0.0000
   0.0630000         14 R         12                   ;resistor, electric
                     21 MUL       12       14          ;thermal power
   0.0011200         31 C         21      -32          ;thermal capac #1
   0.0000
   0.2320000         32 R         31      -33          ;thermal resist #1
   0.0011200         33 C         32      -34          ;thermal capac #2
   0.0000
   0.2320000         34 R         33                   ;thermal resist #2
   1.0000           111 CON
```

Figure 7.8 TUTSIM listing and simulation result of the heat flux of Fig. 7.7.

fluid mechanics. It is further related to solar energy radiation which is a kind of convection of photon gas. We shall give an overview.

Figure 7.9 contains a piece of a fluid pipe with matter flow from left to right. For study and for the working of a Bondgraph we can take two points of view for which we introduce an imaginary observer:

1. Observer travelling with the moving matter, the point of view of Lagrange. Here the control volume travels with the average fluid speed and no fluid penetrates its boundaries (point broken lines of Fig. 7.9). The fluid behaves as a 2-port C-field if we exclude phase changes and chemical reactions (sec. 7.3).

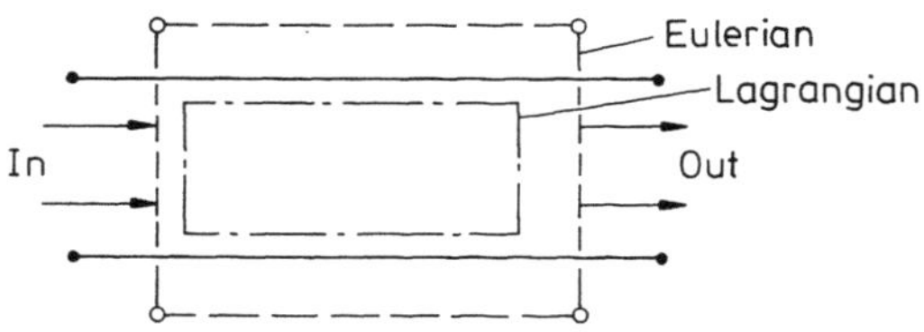

Figure 7.9 Section through a fluid pipe with a control volume for an observer moving with the fluid (Lagrangian frame in broken point lines) and control volume for a fixed observer (Eulerian frame in broken lines). Here fluid enters and leaves and affects the variables in the C-field. Below, total energy, transport is represented by the enthalpy flux $\dot{H}$; further h enthalpy/mass ratio, **u** = internal energy/mass, μ is the chemical potential.

2. Observer fixed to the ground, the Eulerian viewpoint. Here the control volume rests relative to the pipe (broken lines on Fig. 7.9). Therefore matter enters upstream and leaves downstream as denoted by the arrows. So for a reticulation, the arrival and leaving of matter, volume, entropy etc. needs to be taken into account.

Matter or mass flux, whether as fluid, as solid like a moving pushrod, or even as powder carries several flows and the corresponding efforts. This is shown in Fig. 7.10 with the notations of multibonds (a small ring around, sec. 8.1). The multibonds are split up in the center in order to show the normal Bondgraph variables, that is temperature, entropy flow, pressure and volume flow. The component with chemical potential and mass flows is not included for simplicity.

The C-field, representing fluid properties in Lagrangian frame travels with the fluid. The lower part of Fig. 7.10 shows the usual heat engineering and thermal formulae as follows:

- The total energy transport is the enthalpy flux $\dot{H}$, which is the sum of chemical energy and mass flow and of entropy times temperatures.
- Introducing the enthalpy/mass ratio h and the internal energy mass ratio u, the enthalpy transport splits into internal energy part and a hydrostatic part.
- With incompressible fluid, volume flow is the same in and out; so the internal energy part can be neglected and only the hydraulic part retained: This is the basis of fluid power or hydrostatic engineering in chapter 4.

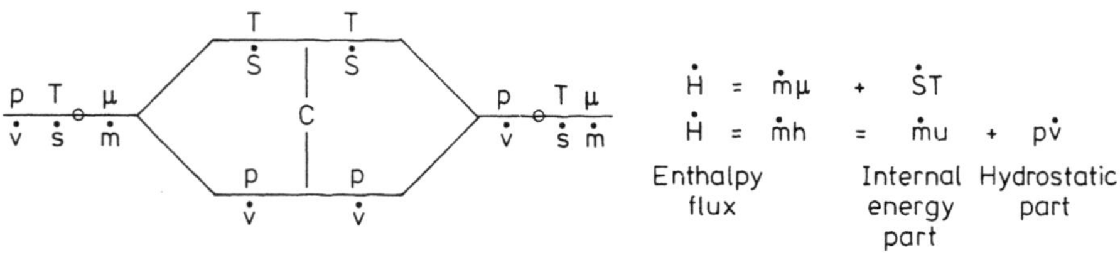

Figure 7.10 Bondgraph for Eulerian frame with multibonds and splitting into components for T/S and p/V. The C-field travels with the fluid.

The internal energy transported by the fluid increases by friction heat (generated entropy), as coming from the hydrostatic resistors of sec. 4.1, but with incompressible fluid it has no influence on volume flow.

If the fluid is only almost incompressible like hydraulic oil and water, the increase of internal energy transported can be measured by thermometers. Assuming that heat conduction (through the walls) out of a machine is negligible, the thermometers can be used to measure the efficiency of water turbines and hydrostatic machines. In thermal power engineering one uses frequently reduced or combined variables, such as pressure divided by the root of (absolute) temperature or an equivalent nondimensional combination. These an be introduced into a Bondgraph by special casuality assignments.

The variables of Fig. 7.10 are not efforts in the usual sense, but have the dimensions of energy per mass. Also they do not control the equilibrium, that is the no flow condition, in the usual way (see below). Figure 7.11 shows a partial Bondgraph, on top for a fluid pipe with its compressibility represented by the C-field. Incoming and outgoing fluids are represented by the p-junctions that transform from the Lagrangian to the Eulerian frame. So the figure shows well the change of volume in the pipe with changing temperature, or the increase of temperature with pressure due to the entropy squeezing of sec. 7.1. It is approximate in the sense that it supposes perfect mixing of the fluid in the pipe.

The lower part of Fig. 7.11 gives a general device, called TEFMA (thermal fluid machine, e.g. a gas or steam turbine). Here the multibond for convection with enthalpy as effort and mass flow as flow splits up into thermal, hydraulic and chemical component bonds; further there is a rotary bond with torque and shaft

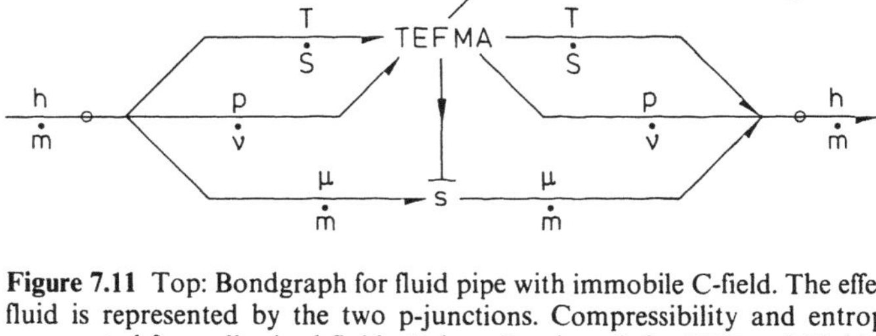

Figure 7.11 Top: Bondgraph for fluid pipe with immobile C-field. The effect of entering and leaving fluid is represented by the two p-junctions. Compressibility and entropy sqeezing are properly represented for well mixed fluids. Below: Bondgraph for ThErmal Fluid MAchine (TEFMA) with detailed bonds. The chemical potential μ has no influence on operation, hence the causality in the vertical bond.

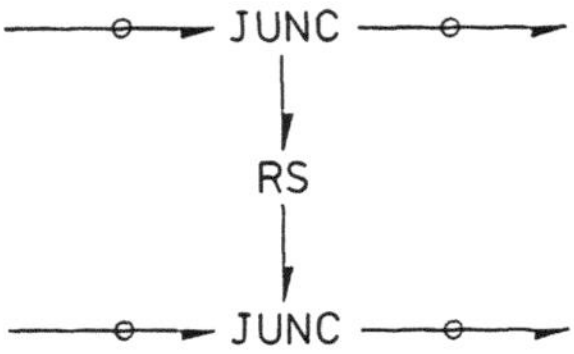

$$\dot{m}_2 = \dot{m}_1$$

$$\dot{m}_4 = \dot{m}_3$$

$$\dot{H}_1 - \dot{H}_2 + \dot{H}_3 - \dot{H}_4 = 0$$

Figure 7.12 HEat eXchAnger (HEXA) with 2 mass flows and a central RS-field for wall conduction. Below: conservation both mass of flows and of overall enthalpy fluxes.

frequency. A thermal conduction bond between the machine and the environment could be added but is usually negligible.

Here only the following can be said in general:

- Mass flow between inlet and outlet is conserved (excluding leakage)
- The shaft power equals the difference of inlet and outlet enthalpy flux (power balance for steady operation).
- The general behaviour of the device or machine is a function of both the fluid property and the design of the machine.
- Applicable causalities depend on how the equations of the fluid machines are presented: Often one uses a combination of variables to reduce complexity. Chemical potential has no influence, hence the causality on the vertical bond.
- Since all machines have friction, including thermal friction, entropy flow will not be conserved but increased.

Similar observations apply to the heat exchanger represented in Fig. 7.12. Here we have two convecting fluids whose mass flows (also volume flows in the incompressible case) can be independently varied. Below there are equations for conservation of mass flow in each pipe and conservation of overall enthalpy flux. We can say that no flow will take place if the pressures are equal between inlets and outlets of each pipe. New entropy is generated, but where it is going to depends on temperature difference and flow directions. This is also true for the direction of entropy flow in the center bonds. In the case of a more compact Bondgraph, the heat exchanger would be called HEXA.

7.2.1 Two Phase Flows and Chemical Reactions

As stated in Fig. 7.11, the chemical potential has no driving force on fluid flow. However, it controls internal equilibrium, say a phase change between steam and water and also possible chemical reactions. This is represented in Fig. 7.13 with

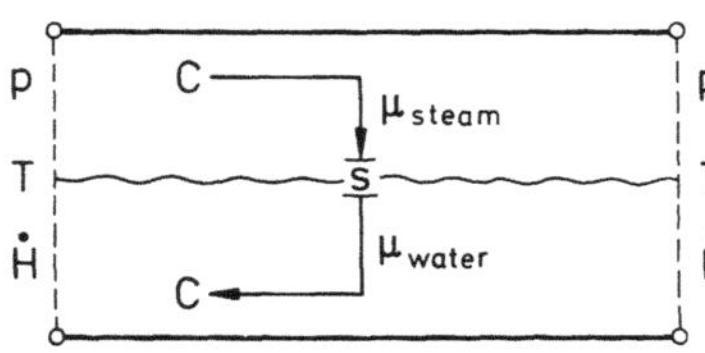

Figure 7.13 Cross-section through a fluid pipe with a supposed interface between 2 phases, steam and water. Here the exchange between the phases is driven by the chemical potential difference.

an assumed interface between water and steam. No exchange will take place if the chemical potentials are equal. Since they depend on variables like pressure and temperature they will change in each device (like the TEFMA of Fig. 7.11). Consequently steam/water exchange can restart in each connecting pipe between devices. Water drops in steam lines are very damaging to steam turbines because of blade erosion.

7.2.2 Capacitor Analogy of Convection

We are envisage a convection bond as a conveyor band with capacitors between a filling and a discharging station as implied in Fig. 7.14. It refers to the hydraulic case with pressure and volume and uses linear capacitors, starting with zero pressure at the filling station. The energy of each capacitor corresponds to the shaded area, which is one half the volume times pressure at the end of filling. So the power becomes one half the pressure times volume flow. We can say there is a convection factor of 0.5.

The convection factor, 0.5 with linear capacitors, depends on the constitutive equations, that is the relation between pressure and volume. If this relation is as shown by the broken line in Fig. 7.14, the factor is higher about 0.8. At the most the convection factor can reach a value of 1 whence convection becomes similar to conduction. If we transport C-fields, as in fluids, the characteristics and convection factor depends on the conditions on the thermal bonds, f.i. adiabatic or isothermal interfacing.

Summarizing convection bonds, the situation is complex and we here gave only the basic physical principles. As stated above, the relations between the variables depend on both the constitutive equations of the fluid and the characteristics of the device, like heat exchangers (HEXA) and thermal fluid machines (TEFMA). So simulation procedures depend on how the characteristics of the devices are given.

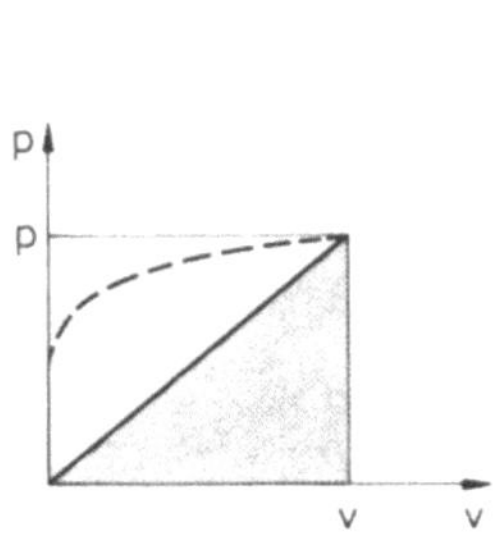

$$E = pV/2$$

$$\dot{n} = \text{capacitors/time}$$

$$\dot{V} = \dot{n} V$$

$$\dot{E} = \dot{n} E = p\dot{V}/2$$

Figure 7.14 Transport of C-fields as model for convection. The graph contains a linear relation between volume and pressure, where the stored energy corresponds to the shaded rectangle and is one half pressure times volume. Hence, convected power is only one half pressure times volume flow. n = frequency of passage of C-fields in a Eulerian frame.

7.2.3 Solar Radiation as Convection

For a concluding example related to the convection of solar energy, in thermal or solar radiation the power or heat flux is not entropy flow times temperature. Figure 7.15 contains a schematic of a thermal radiator (or collector if the power direction is reversed). It has a thermal conducting line at the left, a cavity with black walls in the center and a small hole emitting radiation at the right. Below the equations are written from standard thermodynamics. They start from the radiation pressure, which is one third of energy density u, as follows from Maxwell theory. Note that here we speak about energy/volume ratio in contrast to above, where u denoted energy/mass ratio.

Further the internal energy is proportional to the fourth power of temperature, and entropy density proportional to the third power. Considering the angular distribution, radiated power is proportional to one quarter of energy density times the velocity of light and similarly for entropy radiation. Expressing power by entropy flow and temperature, a factor of 3/4 comes in. So solar radiation is a convection with a convection factor similar to mass flow (Thoma 1980).

Black body radiation, as it is called, had deep influence on the development of physics around 1900 in connection with Plank's radiation formula and quantum theory. He spoke of radiation as a photon gas due to its particular position between conduction and convection.

Naturally entropy cannot be destroyed and must be taken away in solar, thermal and electric collectors, partly by reflected radiation at lower temperature. In the electric case any Carnot formula (sec. 7.1) needs to be corrected for the larger entropy content, leading to a lower efficiency. However, this effect is usually neglegible due to the high temperature (6000 K) of the sun and the many thermal frictions (irreversibilities).

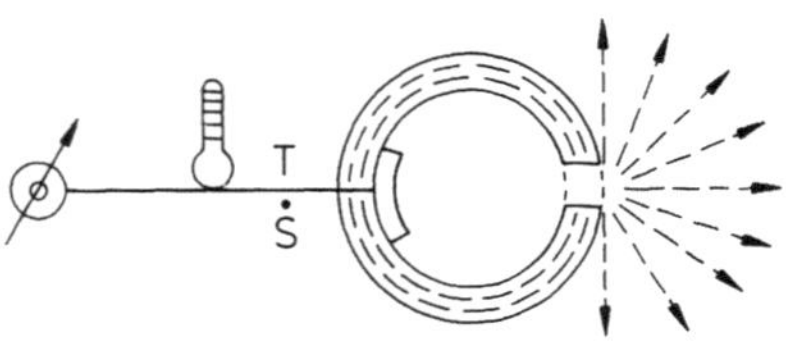

$$p = u/3 \quad \text{(Radiation pressure)}$$

$$u = aT^4$$

$$s = \frac{4}{3} a T^3$$

$$\dot{Q} = \frac{ac}{4} = \frac{acT^4}{4}$$

$$\dot{S} = \frac{as}{4} = \frac{acT^3}{3}$$

$$\dot{Q} = \frac{3}{4} \dot{S} T$$

Figure 7.15 Schematic of solar radiation from a cavity with thermal power supplied from temperature source at left. The equations show that radiated power is 3/4 entropy flow times temperature. Hence radiation is a kind of convection of a photon gas.

7.3 Networks in Chemistry and Physics

Here we shall consider more conceptual, even speculative applications of Bondgraphs to chemical reactions as networks of C- and R- and other elements, followed by some similar ideas in semiconductor physics and electric batteries. To open up, Fig. 7.16 contains an elementary electric circuit with 2 capacitors connected by one resistor. The difference in voltages drives the current which stops if the voltages on the capacitors are equal.

The dissipated power is voltage difference times current, it is always positive and generates entropy according to Fig. 7.4. The sum of the charges on both capacitors is constant. If we make one C-element with variable capacity, f.i. by the movable plates of Fig. 5.8 and vary it periodically, we can shift the electric charge forwards and backwards and each time dissipate energy in the resistor. The energy comes ultimately from the variable attraction force of the plates.

A chemical reaction between a reactant and a product is similar but we have added bonds representing the structural entropy. As shown in Fig. 7.17 the entropy of the disappearing reactant is squeezed out, the entropy of the building products is sucked in. This can be called entropy stripping and is done in the s-junctions and transformers of Fig. 7.17. The difference in the structural entropy is taken from the environment at constant temperature or, in case of thermal insulation, heats/cools the reactants and products. We have the coupling transformers with signal bonds, since they apply no effort in addition to the chemical potential, which alone controls the direction of the reaction and eventual stopping.

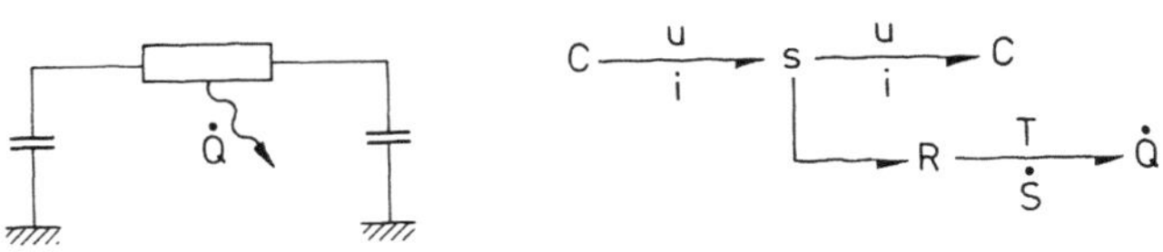

Figure 7.16 Electric circuit with a resistor between 2 capacitor. Current always flows from higher higher to lower voltage; dissipated power equals voltage drop times current.

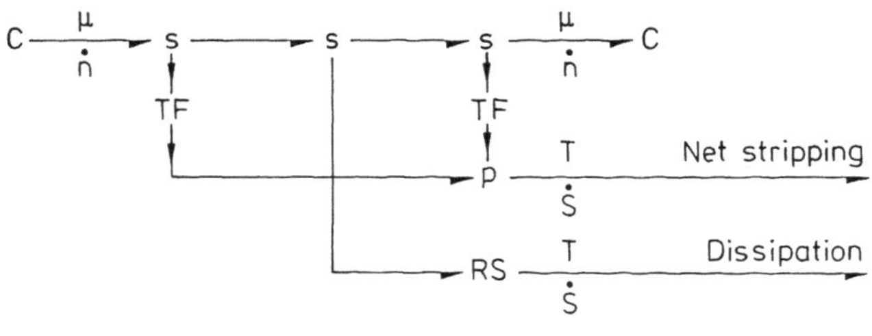

Figure 7.17 Bondgraph for a simple chemical reaction between one reactant, in the C-element on the left, and one product in the C-element on the right. It is similar to Fig. 7.16, but additionally we have stripping of structural entropy that appears reversibly in the environment.

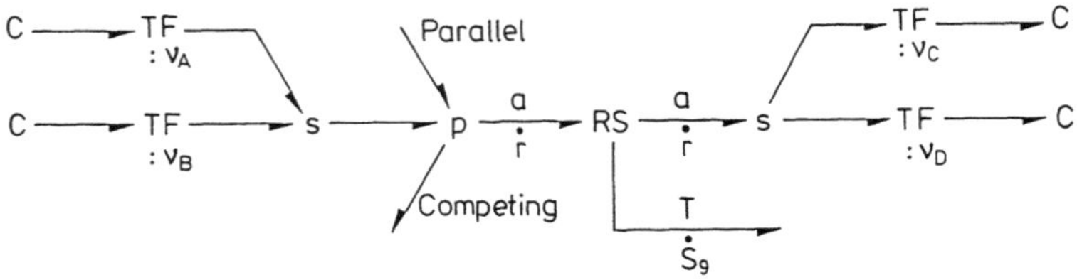

Stoichimetric equations $\nu_A A + \nu_B B = \nu_C C + \nu_D D$
Consfit equa chem pot

Concentration $\mu = \mu_0 (T) + RT \ln \left(\frac{n}{V}\right)$

Figure 7.18 Bondgraph for a chemical reaction with 2 reactants and 2 products, the transformers representing the stoichiometric coefficients and equations below. On the s-junctions the chemical potential is combined to build the net chemical driving force (affinity) a on the RS-field that controls the speed or rate of reaction. The p-junction accepts parallel and supplies competing reactions.

At constant temperature, two very different entropy flows appear:

1. The net stripping of the structural entropy, which is a reversible process.
2. The irreversible dissipation as in the electric circuit of Fig. 7.16.

Coming to chemical reactions, Fig. 7.18 contains 2 reactants and 2 products with additional transformers for the stoichiometric coefficients, according to the reaction equations below. So the chemical potentials combine in the s-junction at the left to give the forward push; this is an effort denoted by an a and and called the affinity. The corresponding flow $\dot{r}$ is called the speed of advancement of the reaction. Similar combinations of the products oppose the advancement of the reaction. The difference attacks the central RS-field. The rate of reaction or chemical kinetics depends on the parameters of the RS-field, the chemical resistance which itself depends strongly on temperature. It is an example of chemical friction mentioned in sec. 7.1. The bonds for stripping of structural entropy are no represented yet exist.

The parallel junction in the middle of Fig. 7.18 indicates possible parallel reactions, witn reactants coming from another reaction and further competing reactions that lead to different products. The art of chemical engineering is to avoid undesirable competing reactions by the proper choice of temperature and pressure. The chemical potential is a simple and fundamental variable associated with mass flow (or molar flow), but details are outside our scope. Figure 7.18 also contains a formula for the chemical potential of gases or substances in weak concentrations; it depends logarithmically on concentration.

Figure 7.19 represents a chemical reaction more in general with multibonds for reactants and products. Instead of an RS-field we have taken a multibond filter that allows some reactions and prevents others in function of temperature. It also produces entropy if a reaction is running.

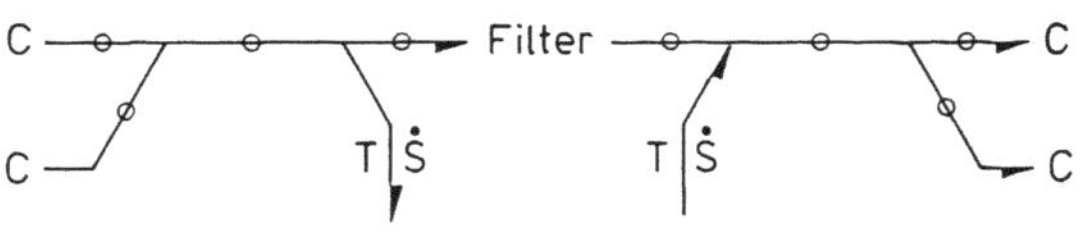

Figure 7.19 Chemical reaction in general with a filter instead of an RS-field in the center; it allows various reactions as function of temperature and pressure.

7.3.1 Osmosis

Osmosis is the penetration of a fluid through very small pores even against an opposing hydrostatic pressure. It plays a great role in chemical membranes and biophysics, f.i. to raise water in trees against gravity. A Bondgraph model then has both hydraulic bonds with hydrostatic pressure and volume flow and chemical bonds with chemical potential and molar flow (Thoma-Atlan 1985). For equilibrium (no flow) chemical potentials must be equal in small pores and pressures must be equal in large pores as in normal hydraulic pipes. This puzzling contradiction leads to a kind of boundary layer on the pore walls with about 25 nm thickness: the boundary layer is pushed by the chemical potential and the central part by hydrostatic pressure difference. So pores of about 50 nm diameter have pressure controlled leakage (leaky membranes); with much larger diameters osmosis disappears.

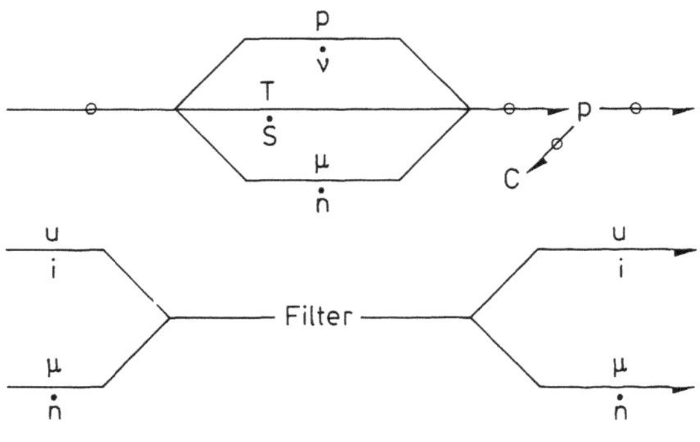

Figure 7.20 Multibond coupled flows in physics, with a coupling C-field between and flow components. Below: Example of a multibond with chemical potential and electric voltage/charge flow or current with a filter in semiconductors.

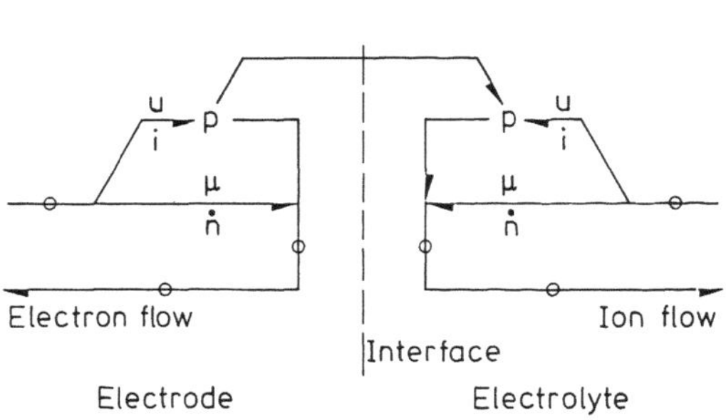

Figure 7.21 Application of Fig. 7.20 to a battery with electrode/electrolyte interface. Due to the filtering effect electrons cannot leave the electrode, ions cannot leave the electrolyte, but electric charge is transferred.

7.3.2 Conceptual Multibonds in Physics

According to the work of G. Falk (Falk 1985) many processes have coupled flows, as shown in Fig. 7.20 with a very stiff coupling capacitor. This includes flows in semi-conductors with chemical and electric bonds appearing in Fig. 7.19 below. There is a filter in the middle that controls the flow in such processes, like in a P/N junction.

Figure 7.21 displays the operation of a battery, in particular the interface between electrode and electrolyte. On the interface between electrode and electrolyte there is a blocking filter (a barrier) to electrons and ions. The electric charge is stripped away (squeezed out) and loaded to the ions in the electrolyte, which is barred by the filter to penetrate the electrode. Consequently both electron and ion flows must return. The reaction is driven by the electro-chemical potential, a combination of chemical potential and voltage. Power comes from a chemical reaction not shown.

We have given here briefly some more speculative Bondgraphs that summarize accepted notions of chemistry and physics in order to inspire the inquisitive reader to further research and study. It is remarkable that analogies, the basis of Bondgraphs, go so far as to allow s-junctions to represent the (stoichiometric) coupling of chemical flows and p-junctions the possibility of parallel and competing reactions. Yet the Bondgraph also expresses the differences and helps to avoid overuse of the analogies.

8 Selected Questions

8.1 Multibonds, Robotics and Circuit Theorems

We treat multibonds and robotic mechanisms together here because robots contain many of them; they are sometimes called vector bonds. They are a group of simple bonds that are somehow related. An example would be the movement of a robot hand in three dimensional space with speeds in the x, y and z directions. We can also think about the transformation into other coordinates, say polar coordinates, and this establishes the connection to mechanisms.

A multibond contains several part bonds or component bonds but we call them strands because the word component has many other meanings. Graphically we represent the multibond by a line, just as a single bond, but with a small ring around it, just like for a real cable a ring (or clamp) keeps the wires together.

Figure 8.1 shows a multi-Bondgraph with separate strands at the left that come together to form multibond 1. The economy in drawing time and space is impressive in large Bondgraphs.

In general both bonds and strands need to be labelled, for which there are the possibilities:

- label bonds by number and strands by letters as done in Fig. 8.1
- label both bonds and strands by numbers.

The first, alphanumeric labelling is more convenient for written out Bondgraphs, especially if the bonds numbers go above 9 as they do in all but very simple cases. Pure numeric labelling is better for theoretical considerations. In both cases we can add a number enclosed in brackets to indicate the number of strands.

Junctions in multi-Bondgraphs couple all strands among themselves. So the central s-junction of Fig. 8.1 couples one strand each of bonds 1 to 3 to equal flows and summing of efforts; a p-junction adds similarly. Box 8.1 contains this as formula. When more selective coupling is required, say between a single strand 2c of bond 2 and bond 4 in Fig. 8.1, the multibond can be split up and the strand equipped with a p-junction. For more complex couplings incidence matrices can be used.

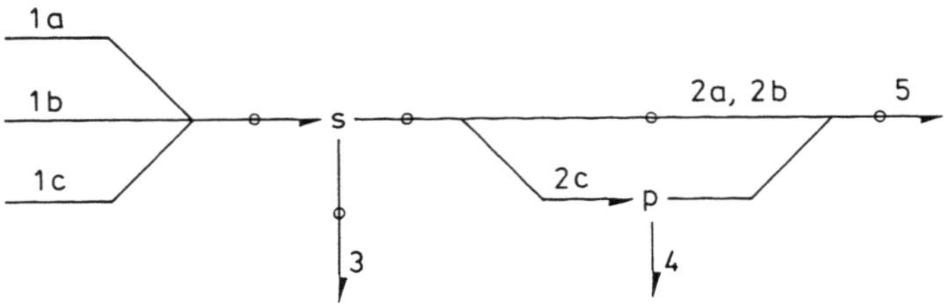

Figure 8.1 Bondgraph with multibonds indicated by a ring around the bond lines. Some multibonds are split up into single strands and labelled 1a, 1b, and 1c. The s-junction connects separately a corresponding strand in all bonds. The parallel junction connects only strand 2c with bond 4 as indicated by the splitting.

Box 8.1 Power and junction equations for multibonds

$$\dot{E}_1 = e_{1a}f_{1a} + e_{1b}f_{1b} + e_{1c}f_{1c} \quad \text{Power}$$

s-junction:

$$f_{1a} = f_{2a} = f_{3a} \qquad e_{1a} + e_{2a} + e_{3a} = 0$$

$$f_{1b} = f_{2b} = f_{3b} \qquad e_{1b} + e_{2b} + e_{3b} = 0$$

$$f_{1c} = f_{2c} = f_{3c} \qquad e_{1c} + e_{2c} + e_{3c} = 0$$

p-junction: permute s and p

Box 8.1 Power and Equations for Junctions in Multibonds.

8.1.1 Modulated Multiport Transformers

Transformers between multiports are the basic Bondgraph elements for mechanisms, especially in robotics. Here the transforming ratios generally depend on a signal like one or several angles; hence we have a multiport modulated transformer to be introduced in the following simple, two-dimensional example.

Figure 8.2 contains a planar mechanism, a lever of variable length rotating around a joint fixed to the ground. The transformer relates length and angle of the lever to the x and y coordinates of the end by the simple trigonometric formulae, which are relations between displacements. Below there is the Bondgraph transformer with the modulating angle from the integrator at the lower left.

Figure 8.3 contains at the top the flow relations (speed and angular frequency) obtained by derivation of the displacement relations in Fig. 8.2. At the top right they are compactly written by the flow transforming matrix $T(\varphi)$. Below appear the effort relations obtained by summing forces written both explicitely and with the effort transforming matrix. Both matrices are seen to be transposes of each other and this is no coincidence.

Multiport transformers are power conserving like all transformers and consequently the effort and flow transforming matrices are the transpose of each

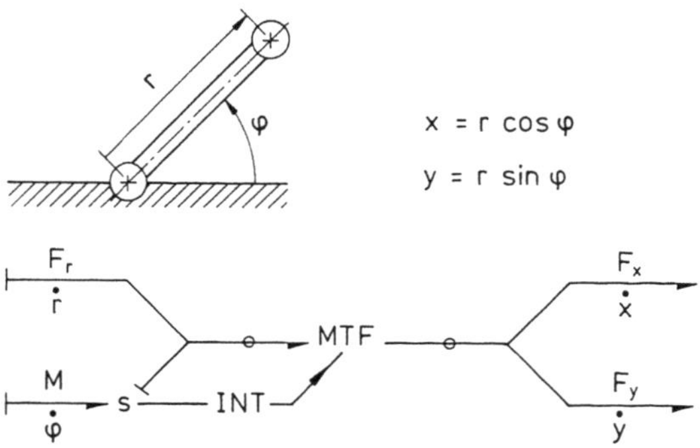

Figure 8.2 Two dimensional mechanism and corresponding multi Bondgraph below. The transformation equations are simple in this causality; they need only one integrator to find the angle.

$$\begin{vmatrix} \dot{x} \\ \dot{y} \end{vmatrix} = \begin{vmatrix} \cos\varphi & -r\sin\varphi \\ \sin\varphi & r\cos\varphi \end{vmatrix} \cdot \begin{vmatrix} \dot{r} \\ \dot{\varphi} \end{vmatrix} \qquad \begin{vmatrix} F_r \\ M \end{vmatrix} = \begin{vmatrix} \cos\varphi & \sin\varphi \\ -r\sin\varphi & r\cos\varphi \end{vmatrix} \cdot \begin{vmatrix} F_x \\ F_y \end{vmatrix}$$

$$\begin{vmatrix} \dot{x} \\ \dot{y} \end{vmatrix} = \begin{vmatrix} T(\varphi) \end{vmatrix} \cdot \begin{vmatrix} \dot{r} \\ \dot{\varphi} \end{vmatrix} \qquad \begin{vmatrix} F_r \\ M \end{vmatrix} = \begin{vmatrix} T^t(\varphi) \end{vmatrix} \cdot \begin{vmatrix} F_x \\ F_y \end{vmatrix}$$

Figure 8.3 Constitutive equations in matrix form for Fig. 8.1, above for flows and below for efforts. The matrices are the transpose to each other by Karnopp's theorem.

other (row and columns interchanged), as Karnopp has shown in 1969 by a short matrix manipulation. See Box 8.2.

We observe the following:

- The remarkable Karnopp theorem allows to determine the effort matrix by transposing to flow matrix which is generally much easier to write from kinematical relations, as was done in Fig. 8.2.
- a mechanism has as many degrees of freedom as its multiport transformer accepts flow input causalities, or as its flow matrix has indepent flow inputs.

Box 8.2 Derivation of Karnopp's Theorem

if $f_{2j} = T_{ji} f_{1i}$ then $e_{1j} = T_{ij} e_{2j}$

where T_{ij} is the transpose T_{ji}

Demo from power conservation
(use Einstein summing convention over repeated indexes)

$e_{2j} f_{2j} = e_{1i} f_{1i}$ powers are equal

$e_{2j} T_{ji} f_{1i} = e_{1i} f_{1i}$ by inserting

for equality $e_{2j} T_{ji} = e_{1i}$ QED

Box 8.2 Derivation of Karnopp's Theorem.

- the causality of Fig. 8.2 was appropriate since the parameters could be expressed as a function of input displacements. Naturally they have to be recomputed at each calculating step in function of the angle φ.
- the inverse causality, that is vertical and horizontal speed inputs is bad since it would in addition require a parameter inversion at each point.

8.1.2 Complex Mechanisms and Transformers

We have explained the essential features here on a simple mechanism with two degrees of freedom. More complex mechanisms, either planar linkages or three dimensional manipulators, can be represented by multiport transformers which are modulated by several angles. They require much computing time and often lead to derivative causalities, f.i. when they connect two inertias. The remedy of element slipping of sec. 6.4 is not feasible with modulated transformers. What remains are well placed compliances of robot arms, which should also help to get the transformer equations into a good causality. For an excellent further treatment take Tiernego-Bos 1985.

Figure 8.4 is a compact representation of a multiport transformer and of the constitutive matrices which are the transposes of each other by Karnopps theorem. As indicated by the signal bond (from below), the parameters can depend one or more parameters and do so in a robot with several degrees of freedom, that is with several joints between the members.

8.1.3 Tellegen's Circuit Theorem

A prominent theorem in electric circuit theory is best explained with Bondgraphs. Take any Bondgraph (Fig. 8.5) and deform it such that all junctions are inside a conceptual boundary, the circle in broken lines on Fig. 8.5. Then label the penetrating bonds to the other elements say from i = 1 to mx calculate the sum of the powers in them. This sum is zero as results from manipulations of the equations of the junctions, independently from the elements outside.

We have three levels of the theorem of zero power sum, called the theorem of Tellegen:

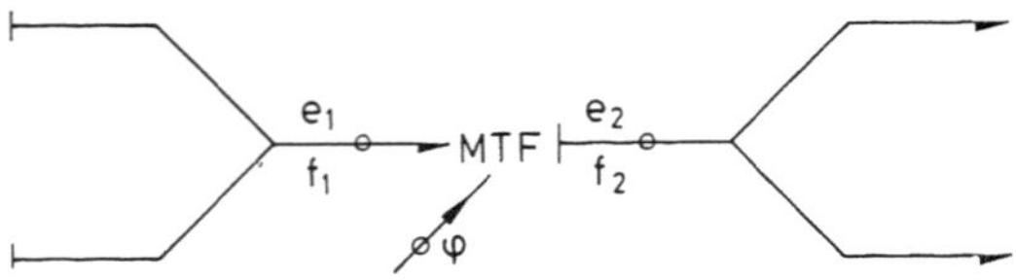

$$\underline{f}_2 = T(\varphi) \cdot \underline{f}_1$$
$$\underline{e}_1 = T^t(\varphi) \cdot \underline{e}_2$$

Figure 8.4 Concise representation of a multiport transformer with constitutive equations for flows and efforts below.

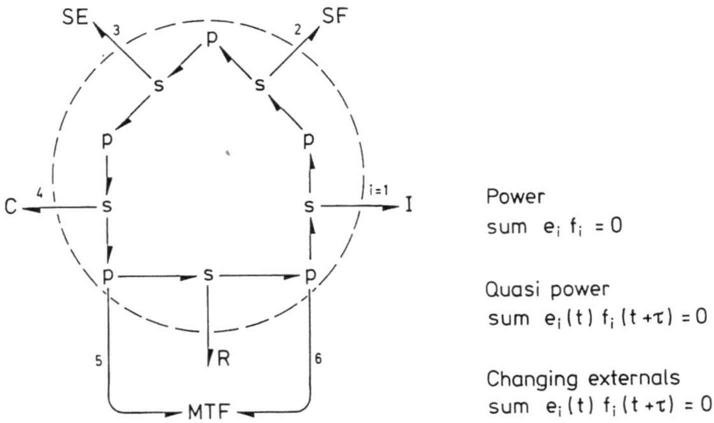

Figure 8.5 Deformed Bondgraph with all junctions inside the circular boundary (broken line) and all other elements outside. The sum of the powers in the penetrating bonds are equal, also the quasi-powers by Tellegen's theorem. This is true also if the external elements change as they do in biological structures.

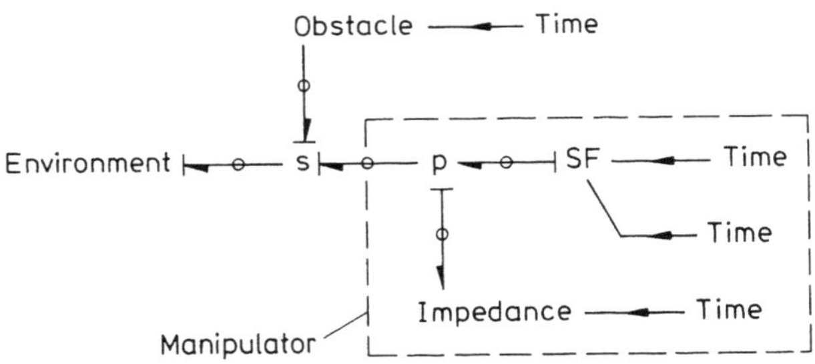

Figure 8.5a Multi-Bondgraph for Hogan's impedance control. The manipulator part is a set of flow sources with impedances in parallel, the environment is in admittance causality and obstacles are coupled in by s-junctions.

1. All effort and flows are measured at the same time so that their product is the real power. That the penetrating powers sum up to zero is pretty obvious since there are only power conserving junctions inside.
2. take efforts and flows at different times, whence their product has no immediate physical significance and is called quasi power. Conservation of quasi power is not obvious.
3. Conservation of quasi power holds also if the outer elements change during observation. So Tellegen's theorem also applies to changing networks which are significant in biology (Atlan, Katchalsky 1973). Condition is that only the junction structure remains.

This and similar circuit theorems may be distant from the needs of the practical simulator, but they give a sight on the great unifying power of Bondgraphs.

8.1.4 Impedance Control for Robots

Whilst on the subject of robotics, Hogan's impedance control method is one of the most promising of the many (Hogan 1985); it is based on Bondgraph ideas. The problem is to control the movement of the work piece around the 3 constituents:

Manipulator, Environment, Obstacles

The key ideas are as follows:

- The manipulator is reticulated as a set of flow sources with a parallel subgroup in impedance causality. This is the so called Norton equivalent circuit as shown on Fig. 2.8 with the flow source.
- The environment is reticulated as a combination of C-, I-, and R-elements in admittance causality; with preponderant inertias this leads to integral causalities. The environment comprises the parts to be moved and/or their contact with external structures.
- Obstacles are impedance subgroups (not path constraints) and coupled in by s-junctions. These subgroups may be moving to represent moving obstacles.

Based on these ideas, many conclusions follow easily. Most elements are time dependent due to manipulator movement, changes in obstacles and disturbing forces on the work pieces. Such disturbances are mainly inertial and arise when a screw or bolt should be assembled into a hole or more in general from desirable contact with external structures.

The subgroups and elements form causal loops and can become unstable with non-passive components. Stability depends on positions and other variations, but can be checked by the methods of secs. 6.1 and 6.7.

Figure 8.5a gives an overview of impedance control with a multi-Bondgraph. The manipulator part comprises a set of flow sources and parallel impedances, the environment is represented in admittance causality (effort in, flow out) and the obstacles in impedance causality (flow in, effort out).

Further research and development based on the concept of impedance control will advance the art of robotics. It is good to see how the universal power and beauty of Bondgraphs is also effective in this difficult field.

8.2 Bondgraphs for Elements Distributed in Space

Many systems have dynamic properties distributed in space, not concentrated as usually assumed; they are described PDE (partial differential equations). Examples are fluid pipes with pressure waves, vibrating bars and beams and electric wires. PDE are required if the wave length of the disturbance is comparable to the length of the pipe, beam or wire. For lower frequencies concentrated elements are normally sufficient. The velocity of propagation of the disturbance, that is the velocity of sound in mechanical systems depends on the medium: 330 m/s in air, 1310 m/s in fluids, 5500 m/s in steel bars (less for bending oscillations). Wave length is of course velocity of sound divided by frequency.

The problem is to model and simulate distributed elements efficiently. There exist two different methods, which we shall explain in the example of a fluid pipe.

1. The segment method: Divide the pipe into segments or slices and give each a mass, resistance and inertia and combine the whole into a Bondgraph.
2. The modal method: With simple geometry, such as a straight pipe, solve the PDE to obtain a set of Eigenfunctions. Each is represented by some I,C,R-elements and transformers in a kind of resonant circuit. All are combined into a Bondgraph.

The segment method is conceptually simple and shown in two variants on Fig. 8.6; they differ essentially in the causalities for connection to the external subsystem. We have at the top two I- and R-elements and three C-elements, as suitable for flow input causality. Below we have three I- and R-elements and two C-elements for pressure input causality. Naturally finer divisions, that is more segments, are possible. The parameters for each segment have to be divided from the total inductance, resistance and capacitance of the pipe and depend on the number of segments as indicated in Fig. 8.6. Consequently the resonance frequency of each segment increases and its damping decreases (compare Figs. 6.7, 6.8) with more segments. Further, the required time step decreases and this limits the practicable number of segments.

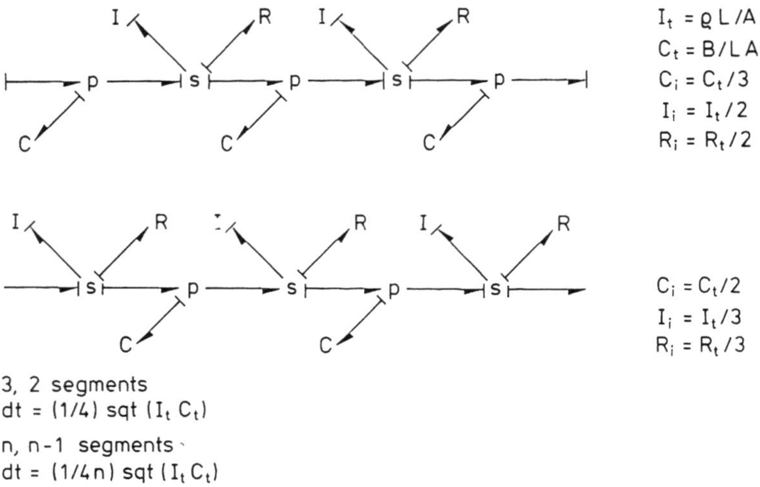

Figure 8.6 Reticulation of a distributed element system, here a fluid pipe by the segment method with 3 and 4 elements. The upper part is suitable for flow input causality and the lower one for pressure input. Below the required time steps appear: they decrease with more segments.

An advantage of the method is that the segments are located in space and can be given different parameters for non-uniform pipes. A weakness seams to be that with more than about 8 segments very high and weakly damped Eigenvalues appear, much higher and weaker then expected from the dynacheck of sec. 6.4. Consequently the simulation will contain quasi permanent oscillations of very high frequency, which are unrealistic. This means that they are not observed in practice, the why depends on further research, perhaps additional resistances not in the model.

For a comparison of segment and modal methods see Thoma–Richter 1987 and also for the details of the following summary of the modal method.

The modal method works on the PDE for either pressure or flow (accoustic wave equation), which are obtained from a momentum balance and continuity equation. There are added terms for injected pressure and flows. We can solve the pressure equation, a procedure suitable for flow input causality, or the flow equation for pressure input causality.

For solving the flow equation in Fig. 8.7, we set the injected flow to zero and assume a boundary condition of zero pressure at both pipe ends. Proceeding along classical lines of mathematical physics, we obtain the flow as a sum of space dependent terms with the cosine function, the so-called Eigenfunctions. Each is multiplied by a time dependent term. The external pressure is applied at points x1 and x2, and the flow at these points calculated by the sum of the Eigenfunctions multiplied by the time dependent terms.

As shown in Fig. 8.7 below, each Eigenfunction can be represented by a series circuit. It is excited by the two transformers, each having as parameter the magnitude of the Eigenfunction at the point of connection with the external system, that is x1 and x2. To underline this important result:

1. The flow equations of the transformer shows how the time dependent partial flows q_i combine to make the overall flow according to Fig. 8.7 at the top.
2. The effort (pressure) equations of the transformers show the resonance of the modal flows χ are excited by the external pressure.

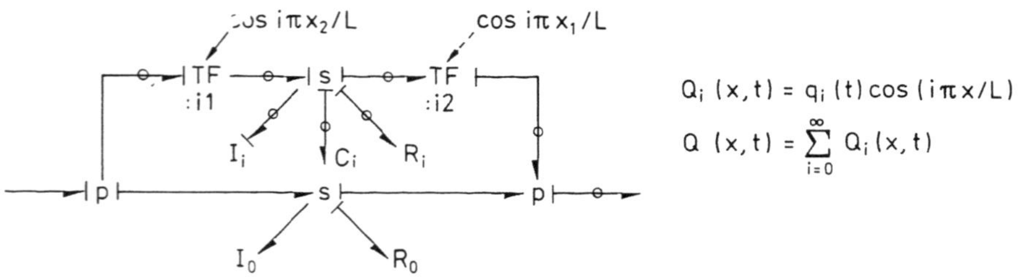

Figure 8.7 Modal reticulation, at the top the formula for the total flow Q(t) as a sum of partial flows Qi, which are a product of the time dependent modal flows χ and a space dependent mode shape or Eigenfunction. Below: Representation by a Bondgraph with a mode stack. The modal flows are coupled by the transformers to the partial flows and the parameters come from the space dependent part. The zero mode is written separately and contains total inductance and resistance.

If the injections x1 and x2 happen to be at zero point of the cosine function, the particular resonance mode will not be excited at all. This is an important result in the theory of vibrating structures with distributed elements.

The lowest mode of Fig. 8.7 contains no C-element, only the inductance I_0 and resistance R_0 of the whole pipe. This is the zero mode above which a mode stack with many modes appears, each having a higher resonance frequency. The key idea is to neglect all higher modes and to retain only those within the frequency range of interest: in practice a stack of two or three modes.

Figure 8.8 shows the parameters belonging to each mode in terms of the total inductance, resistance and capacitance It, Rt and Ct of the whole pipe. Here inductance is independent of the mode number, capacitance decreases with its square. Consequently resonance frequency is proportional to mode number. Further as shown (using the velocity of sound a) the first mode has a wavelength of twice the pipe length, the second equal to the pipe length and so forth, as might be expected.

A possible refinement is to replace all neglected modes by a residual capacitance because a series circuit excited far below its resonance frequency behaves like a capacitor. These capacitances can be summed and should improve accuracy of the simulation, but they require mixed causality and give clumsy TUTSIM code.

Figure 8.9 shows a simplification with only two retained modes and injecting points at the beginning (x2 = 0) and the end (x1 = 1) of the pipe. The transformer parameters simply become + 1 and − 1; a transformer with a parameter of + 1 is an identity and could be omitted.

The modal method is elegant and was contained (for a vibrating beam) in Karnopp Rosenberg (1968), including the idea of residual capacitance. More recent is the survey of Margolis (1985) and the work of Lebrun (1985), who succeeded to include the resistance term in the PDE. The relation to other engineering areas like vibrations theory, Fouier analysis and modal analysis is evident.

Summarizing the above approach: by solving the flow equation we obtain a zero mode with the total resistance and inductance; further a parallel mode stack containing the higher modes, corresponding to the harmonics in Fourier analysis. Each mode contains an I,R,C combination (connected in series); an electric equivalent circuit would be easy to write.

$$I_0 = I_t$$
$$R_0 = R_t$$
$$I_i = I_t/2$$
$$R_i = R_t/2$$
$$C_i = C \times 2/(\pi^2 i^2)$$

frequency

$$f_i = \frac{i}{2L}(B/\varrho)^{0.5} = \frac{ia}{2L}$$

velocity of sound

$$a = (B/\varrho)^{0.5}$$

wavelength

$$\lambda_i = a/f_i = 2L/i$$

Figure 8.8 Formulae for the parameters of the modestack. In zero mode inductance and resistance are equal to total inductance and resistance of the pipe, but for higher modes capacitances decrease. The resonance frequencies correspond to a wavelength of twice, once or a fraction of the length of the pipe.

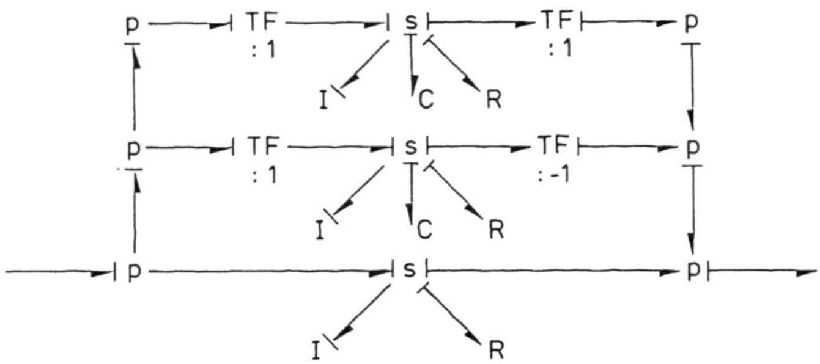

Figure 8.9 Simplification for only two higher modes connected to the rest of the system on both ends. The transformers with the parameter of one (value of the cosine function at x = 0) are identities and can be left out (done on the next figure).

Solving instead the pressure equation for flow input causality one obtains a series mode stack and a zero mode with a total capacity in parallel with the total resistors. The higher modes are in series and each is a parallel resonance circuit. The solution is much less intuitive and no electric equivalent circuit can be written.

8.2.1 Simulation of Hydraulic Ram Pump with Modestack

As an example of a pipe reticulation with a stack of two parallel modes, Fig. 8.10 shows a hydraulic ram. Water comes from the flow source SF11 at the right, with pressure limited by the relief valve R13. It accelerates and runs through the main pipe, called shock tube, and escapes through the valve R25, the ram valve being open at low flows. With sufficient water speed the ram valve closes suddenly, a pressure shock develops that opens the check valve R28 and delivers some water to a high reservoir. When the pressure shock has ended, the ram valve re-opens, the check valve closes and the acceleration of the fluid starts again. We thus have a hydraulic oscillator.

Hydraulic ram pumps of the kind of Fig. 8.10 are useful to take water from a flowing river and nump it to domestic tanks. They were already proposed by the brothers Montgolfier, who are known for their aerostatic balloons, about 200 years ago. In the 19th century they were much used for the water supply to houses and to the construction sites of railways, until superseded by pumps driven by electric motors or gasoline engines. Today they show new applications as well pumps in desert (very dry) countries, where the driving water flow can come from solar electric power and a suitable low flow, low pressure pump.

The lower part of Fig. 8.10 contains the Bondgraph with C-elements C14 and C24 added for causality. The shock tube itself, in broken lines, contains a zero mode with I220 and R230 and two modes above. As mentioned already near Fig. 8.9, unnecessary transformers have been omitted and there only remains one transformer with a parameter of -1 at the right of the first mode.

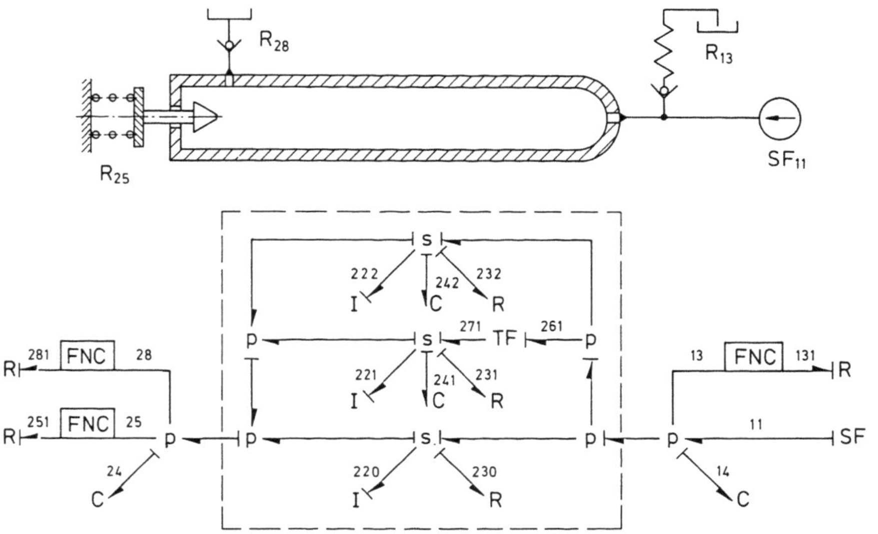

Figure 8.10 Hydraulic ram pump with shock tube, represented by a zero and two higher modes and various valves. It is a hydraulic oscillator that periodically delivers water to the high pressure tank through R28.

Figure 8.11 shows the simulation result in continuous oscillatory mode with a frequency of about 34 Hz. The upper part refers to the shock tube with the mode stack and gives almost square pulses of p24, the output pressure driving the water into the higher tanks of R28. The input pressure p14 tends to rise a little later, about 3.5 msec. after closing the ram valve.

The lower part of Fig. 8.11 shows the simulation of the ram pump where the pipe is reticulated by the segment method (Fig. 8.6 above) with 4 C-elements and 3 I- and R-elements. Comparing both reticulations, flow Q25 and oscillation frequencies are apnroximately equal, but the pulse p24 is much better with the mode stack. It is more nearly-rectangular as required from the theory of the (Joukowsky) shock waves. Higher modes in the mode stack would essentially tend to flatten out the pressure ripples on top of the rectangle.

Figure 8.12 contains the listing for the ram pump including the shock tube with a mode stack and realistic parameters. They represent a tube of 4.5 m length, 60 mm diameter and a ram valve closing at a fluid velocity of 6 m/sec with 0.15 MPa pressure drop. The simulation step is determined by the time constant C24 with R28, neither by the highest mode nor the smallest segment. The values of C14 and C24 have been assumed, somewhat arbitrarily, as 18% of the total capacitance of the pipe. This results from some experimentation with the computer.

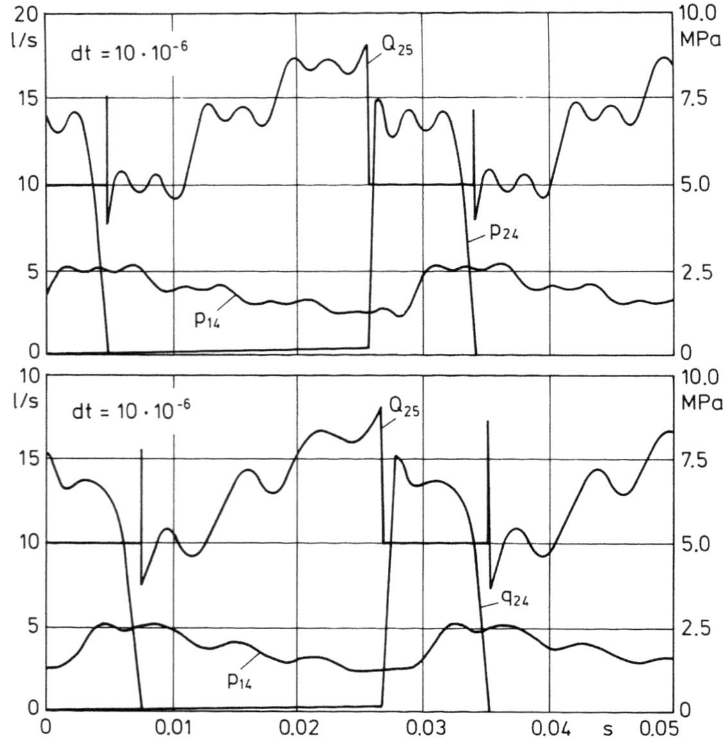

Figure 8.11 Simulation results of the hydraulic ram pump. Top with the mode stack from Fig. 8.10, below with segment reticulation of the same pipe. The modal reticulation represents the shock wave of p24 much better.

8.3 Shocks, Collisions and Hysteresis

We treat here the simulation of special effects of some difficulty but of frequent occurrence.

Shocks happen when two machine elements come in contact and transmit forces (exchange momentum) as long as they stay together. In principle, they can be modelled as shown in Figs. 2.27 and 2.28. This is often sufficient but leads to a rebounce (almost) equal to the impact velocity. It neglects the greatly increased friction on contact which reduces rebounce.

Measuring the relative rebounce is a quality control procedure for steel, especially for ball bearings that should rebounce to about 90%. So a more precise model of impacts is needed here and also in hammers for tunnel construction and drilling of petroleum wells. The following reticulation was developed for such hammers.

```
Timing:  10.000E-06   ,DELTA  ;    0.0510000   ,RANGE
PlotBlocks and Scales:
Format:
     BlockNo,  Plot-MINimum,   Plot-MAXimum;  Comment
Horz:      0 ,   0.0000      ,   0.0500000  ; Time
  Y1:     24 ,   0.0000      ,  10.0000     ; end capac, left
  Y2:     25 , -20.000E+03   ,  20.000E+03  ;
  Y3:     14 ,   0.0000      ,  10.0000     ; end capac, right
  Y4:        ,               ,              ;

   0.0000          1 CON                           ;ARIete with modes
   0.0000          2 CON                           ;driven by SF11 + R13 byp
  16.000E+03      11 SF      111                   ;hydraul source
                  13 FNC     131
     1    0.0000        0.0000
     2    1.0000        0.0000
     3    2.0000       16.000E+03
   1.3000         14 C        11     -13    -220   ;end capac, right
                            -261    -222
   1.7131
   1.3000         24 C       220     -25     221   ;end capac, left
                             222     -28
   6.9674
                  25 FNC     251
     1   -0.1500000    -16.000E+03
     2    0.0000        0.0000
     3    0.1500000    16.000E+03
     4    0.2000000     0.0000
     5    0.3000000     0.0000
                  28 FNC     281
     1    0.0000        0.0000
     2    6.0000        0.0000
     3    8.0000       10.000E+03
   1.2000        111 CON                           ;file bonsi82.sim
   1.0000        131 R        14                   ;relief valve
   1.650E-06     220 I        14     -24    -230   ;zero mode induct
   4.374E+03
 830.000E-09     221 I       271    -231    -241   ;#1 mode induct
                             -24
  -1.218E+03
 830.000E-09     222 I        14     -24    -232   ;#2 mode induct
                            -242
   1.011E+03
   7.200E-06     230 R       220                   ;zero mode resist
   3.600E-06     231 R       221                   ;#1 mode resist
   3.600E-06     232 R       222                   ;#2 mode resist
   1.5000        241 C       221                   ;#1 mode capac
 -13.3410
   0.3700000     242 C       222                   ;#2 mode capac
  -0.2670190
   1.0000        251 R        24                   ;ram valve
  -1.0000        261 TF      221
  -1.0000        271 TF       14
   1.0000        281 R        24                   ;deliv check valve
                 999 REM
;without x87 coproc slightly diffe results
```

Figure 8.12 TUTSIM listing of the simulation of the hydraulic ram with mode stack. The parameters refer to a shock tube of 4.5 m length and ram valve R28 closing at 6 m/sec fluid velocity at 0.15 MPa pressure drop.

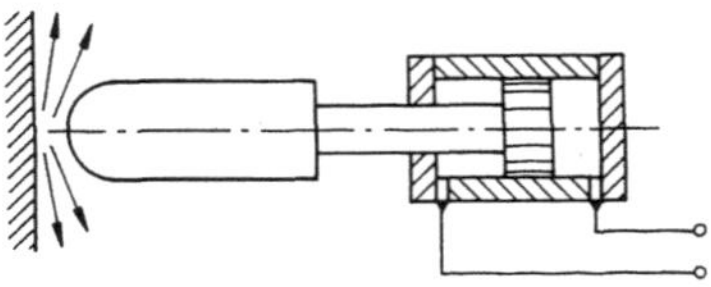

Figure 8.13 Horizontal hammer as example of impact in the soil at the left; the task is to simulate contact forces, penetration and any rebounce.

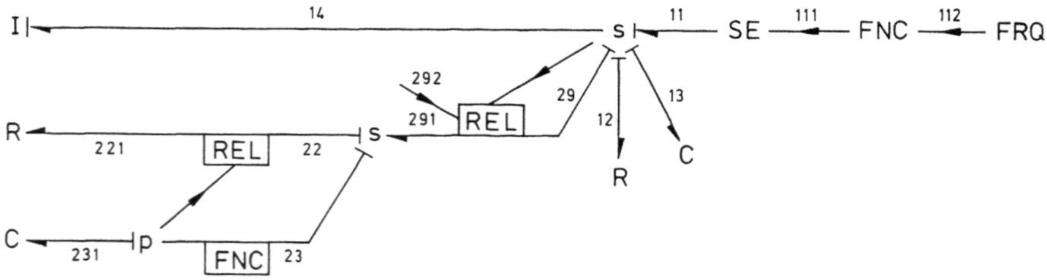

Figure 8.14 Bondgraph for a hammer impact elastic forces from C23, friction force from R221; it is switched on by relay 22 when the hammer is beyond the soil face. The optional relay 29 switches the contact force off when the hammer runs backwards to prevent sticking.

To fix ideas, Fig. 8.13 shows a horizontal hammer driven hydraulically by some periodic pressure or force source. The hammer impacts the earth wall to the left. The corresponding Bondgraph in Fig. 8.14 has a force source SE11, hammer mass I14, return spring C13 and air resistance R12. The block FRQ112 makes a sinusoidal force which is rectified (only positive forces remain) by FNC111.

The following elements model the hammer soil impact:

- The non-linear spring C23 (C231 and FNC23) gives the elastic properties of the soil, similarly to Fig. 2.28, including the position output of Fig. 2.27.
- The soil friction R221 is switched on by the relay REL22 on contact (when x231 is ca. 4 cm).
- On withdrawing the hammer, some soils do not hold it back. So relay RE29 switches off the soil force when the hammer speed is negative.

Figure 8.15 is the TUTSIM listing corresponding to the Bondgraph of Fig. 8.14. This is a comprehensive model for different soils which could be simplified as follows:

- If the hammer sticks in the soil, we could eliminate REL29 (replace it by a gain of one) or put bonds 22 and 23 directly to the upper s-junction.
- With REL29 retained we could leave out R22, since the relay itself eliminates any rebounce. But if retained, R22 serves to adapt to the simulation of experimental data for the contact force on impact.

In fact, C23 gives a force proportional to penetration and R22 to speed, so by adaption many experimental curves can be reproduced. With rebounce the

```
Timing: 100.000E-06  ,DELTA  ;    0.2010000  ,RANGE
PlotBlocks and Scales:
Format:
     BlockNo,  Plot-MINimum,    Plot-MAXimum;  Comment
Horz:      0 ,    0.0000      ,    0.2000000  ; Time
  Y1:     14 ,-500.0000       ,  500.0000     ; Hammer Mass
  Y2:     29 ,-100.0000       ,  300.0000     ; Force Switch
  Y3:    231 ,   -2.5000      ,    7.5000     ; soil elasticity
  Y4:        ,                ,               ;


  80.0000           11 SE     111                          ;Force Source
   0.0100000        12 R       14                          ;Air Friction
   0.2000000        13 C       14                          ;Return Spring
   0.0000
   0.0064000        14 I       11      -12      -13        ;Hammer Mass
                              -29
   0.0000
   4.0000           22 REL    221      221      222        ;Friction Switch
                              231
                    23 FNC    231
     1     0.0000          0.0000
     2     4.0000          0.0000
     3    10.0000          1.400E+03
   0.0000           29 REL    291      291      292        ;Force Switch
                               14
                   111 FNC    112
     1   -10.0000          0.0000
     2     0.0000          0.0000
     3    10.0000         10.0000
  10.0000          112 FRQ                                 ;Periodic Excitation
   1.0000
   1.0000          221 R       14                          ;Soil Friction
   0.0000          222 CON
   1.0000          231 C       14                          ;soil elasticity
   0.0000
                   291 SUM     22       23                 ;sum o forces
   0.0000          292 CON                                 ;file bonsi83a.sim
```

Figure 8.15 TUTSIM listing corresponding to Fig. 8.14, with the comprehensive soil model. The parameters correspond to the compact mechanical units of sec. 6.6.

REL29 is not needed and the rebounce ratio follows from the relative values of C23 and R22. This ratio can be calculated from the formulae of sec. 6.4 (Fig. 6.7 and Fig. 6.8), but it is often easily found by different simulation runs. Comparing calculated rebounce with computed rebounce is a good student exercise.

Figure 8.16 contuins the simulation; at the top without the relay REL29, whence the contact force seems to be negative for a moment on withdrawal until REL22 switches it off when leaving the soil. In the lower part of the figure REL29 is included and the impact force is always positive.

The model of Figs. 8.14 and 8.16 can be used to model many other collisions or impacts, including the effect of backlash in gears, with variable, positive and negative torque. Here we would need to make two impact faces (positions), when the gears contact one or the other end of their free travel. Naturally the simulation time step needs to be watched; it corresponds either to the resonance frequency of I14 and C23 or to the time constant of R22 and C23, whichever is less. The harder the impact, the harder the computer has to work for the simulation, in other words the shorter the time step required.

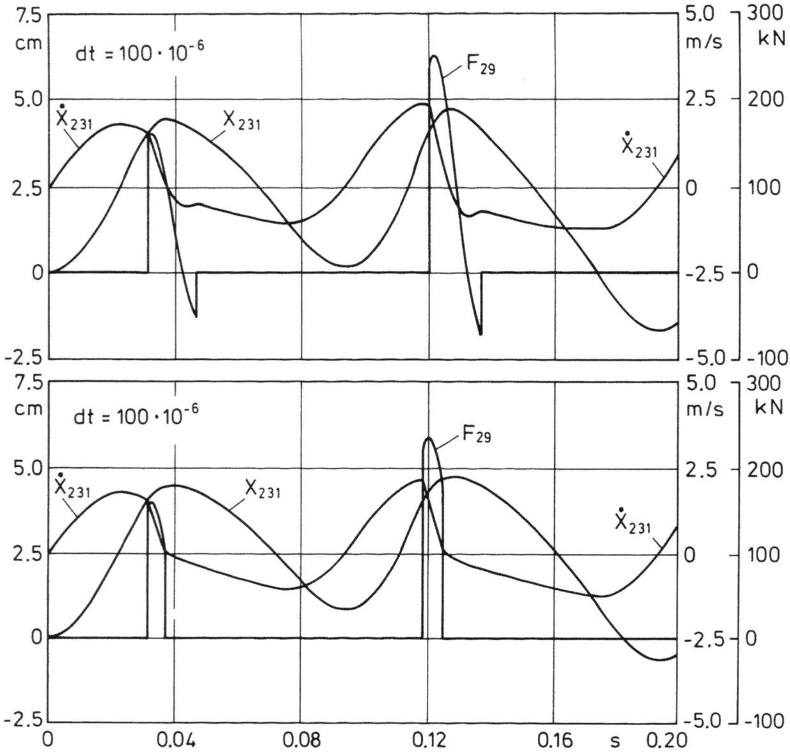

Figure 8.16 Simulation result of hammer impact: On top without relay 29 and with momentarily negative force (F29), hammer speed ẋ 14 and hammer position x 231. On bottom the same with relay 29 that prevents negative contact (sticking) force.

8.3.1 Hysteresis

Magnetic hysteresis is well known from the B/H curves in ferromagnetic material, but mechanical examples also exist. For integral causality we need to compute the current of an inductor as a function of voltage, that is the H-field as a function of the B-field (the inverse of the usual). This is mandatory, not only desirable, since there is only one H value for each B value, but several B values for each H value, particularly a smaller B when H decreases.

Bondgraphing can handle this difficult situation by a relay that switches between the two H(B) curves depending on the time derivative of B.

Figure 8.17 contains at the top an electric resonance circuit, where the inductor with hysteresis is indicated by the usual block with the I-element. Below the hysteresis is enlarged to show relay REL16 switching between the forward and backward characteristics of FNC14 and FND151 (with GAI15); the FND block doubles FNC14, resulting in the symmetric characteristics in Fig. 8.18.

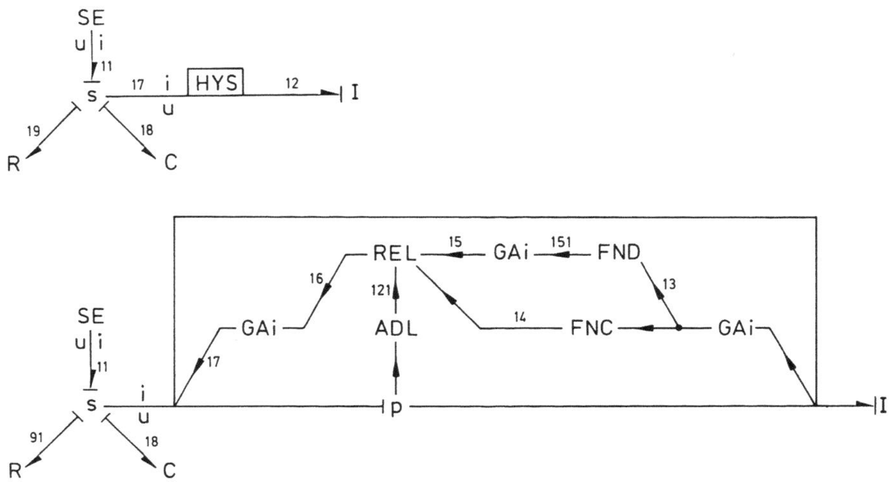

Figure 8.17 Electric resonance circuit where the inductor has hysteresis, top as block-Bondgraph. Below detail with magnetic characteristic generated in FNC14 and FND151; they are switched by relay REL16 depending on the rate of change of voltage.

Instead we could use a separate FNC15; this allows more liberty but requires more programing points.

The ADL block (algebraic delay) is needed to break the algebraic loop built with the connections 14 and 15.

Figure 8.18 contains the simulated magnetization curves, calculated as H-field in function of the B-field (current in function of voltage) but printed in the usual way (B-field vertical, H-field horizontal). The characteristic is obtained by starting the resonance circuit with a constant voltage from the source. The induced oscillations will decay and also write the horizontal parts. For this signal 16 appears on the x axis instead of the usual time.

Below there is oscillation as a function of time of the resonance circuit. The damping increases at lower amplitude because then the hysteresis absorbes more relative energy per cvale. This energy increases with the relative width between both parts of the H/B characteristics.

Figure 8.19 contains the corresponding TUTSIM listing. It runs without difficulties. Relay chatter or other trouble could be expected, however, since the horizontal part characteristics represents an infinite gain when the relay switches between forward and backward characteristics.

8.4 Analogies, Linear Graphs and Portality

We return here to the question of analogies in mechanical systems, namely, which variables are to be taken as efforts or flows. In our book and most Bondgraph publications force is taken as effort and speed as flow, and similarly torque and

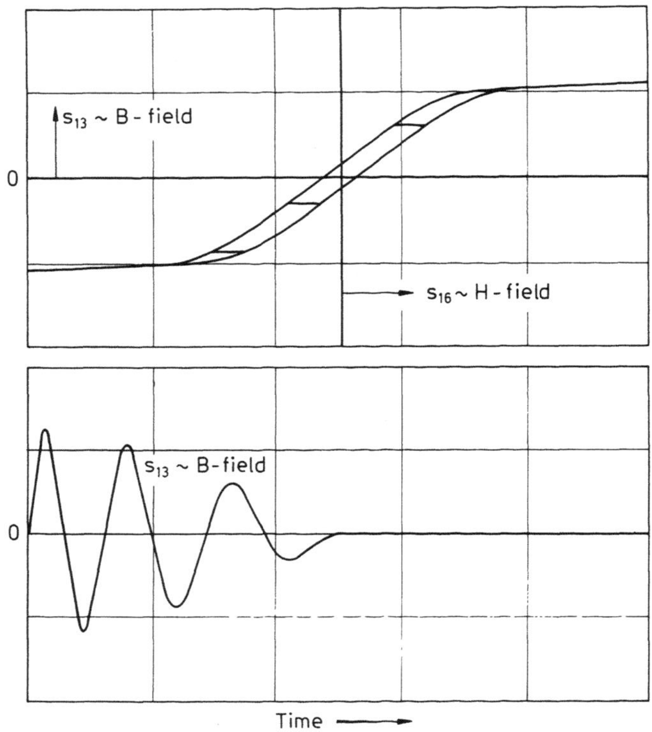

Figure 8.18 Oscillation of the circuit with hysteresis, at the top with the current/voltage characteristic (corresponding to the B/H characteristic of the magnetic field). The horizontal parts come from switching between the forward and backward characteristics on inversion of the voltage rate. They imply infinite gain and can lead to trouble (relay chatter). Below the current as a function of time. The oscillation is more damped at lower amplitude.

rotation frequency in mechanical rotary systems. This is the force-effort analogy. However, there also exists the dual (inverse), the force flow analogy with force as flow and speed as effort. It is always used with another graphic system representation, the linear graphs.

Figure 8.20 shows the changes needed for using the force flow analogy, including that momentum and displacements are interchanged. Thus a better terminology would be effort integral for the real displacement and flow integral for the real momentum. Generally, the Bondgraph terminology is adapted to the force effort analogy, but the force flow analogy has some advantage coming from the notion of through and across variables (see Fig. 8.22 below).

Figure 8.21 shows some consequences of the force flow analogy: springs become I-elements, masses C-elements, hydrostatic machines gyrators and electric motors transformers. This last point is true for class transformers, whilst impedance transformers remain as they are. Such transformers connect variables

```
Timing:   0.0500000  ,DELTA  ;  40.1000      ,RANGE
PlotBlocks and Scales:
Format:
     BlockNo,  Plot-MINimum,  Plot-MAXimum;  Comment
Horz:    16 ,  -2.0000     ,    2.0000     ; switches B/H curves
  Y1:    13 ,  -2.0000     ,    2.0000     ; scale for sizes
  Y2:       ,              ,               ;
  Y3:       ,              ,               ;
  Y4:       ,              ,               ;

   1.5000          11 SE      111                   ;driving voltage
   1.0000          12 I        11     -18     -19   ;integra in magne flux
   0.0000
   1.0000          13 GAI      12                   ;scale for sizes
                   14 FNC      13
     1   -1.0000           -1.2000
     2   -0.9800000        -1.0000
     3   -0.9300000        -0.9000000
     4   -0.8600000        -0.8000000
     5   -0.7700000        -0.7000000
     6   -0.5300000        -0.5000000
     7   -0.4000000        -0.4000000
     8    0.1500000         0.0000
     9    0.6700000         0.4000000
    10    0.8100000         0.5000000
    11    0.9300000         0.7000000
    12    0.9800000         0.8000000
    13    0.9900000         0.9000000
    14    1.0000            1.0000
    15    1.0200            1.2000
  -1.0000          15 GAI     151                   ;invert sign going down
   0.0000          16 REL      15      15      14   ;switches B/H curves
                              121
   1.0000          17 GAI      16                   ;scale for sizes
   1.0000          18 C        17                   ;circuit capacitor
   0.0000
   0.0000          19 R        17                   ;wire resistance
   1.0000         111 CON                           ;file bonsi83b.sim
   0.0000         121 ADL      11     -18     -19   ;break alge loop t REL
  14.0000         151 FND     -13
```

Figure 8.19 TUTSIM listing of the resonance circuit with magnetic hysteresis in Figs. 8.17 and 8.18.

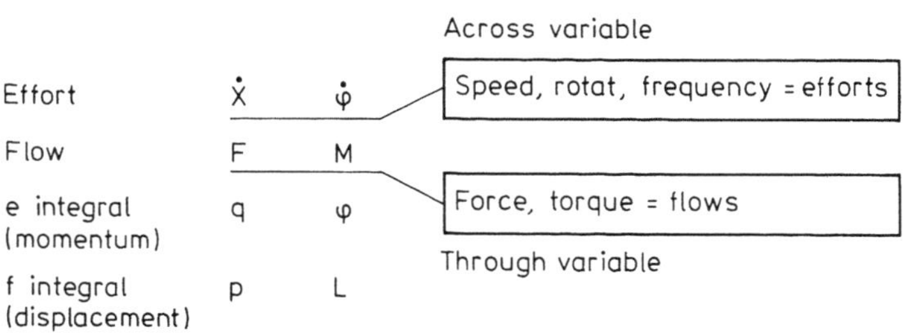

Figure 8.20 Force flow analogy where speed and rotation frequecies are efforts and forces and torques are flows. Real displacement becomes effort integral and real momentum becomes flow integral. With this analogy all efforts are cross variables and all flows through variables.

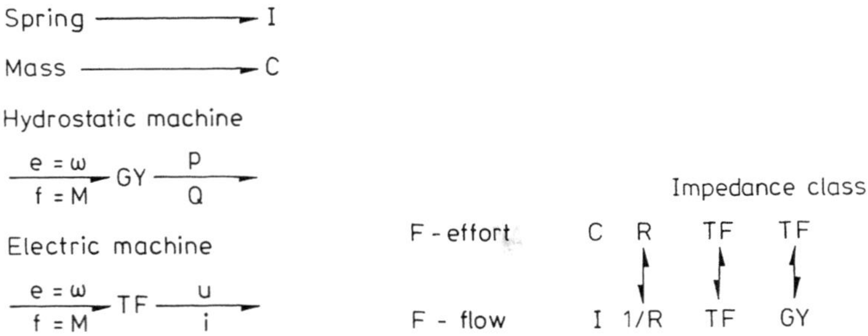

Figure 8.21 In the force flow analogy a spring is an I-element, a mass a C-element and hydrostatic and electric machines turn into gyrators and transformers. Below: C- and I-elements interchange, impedance transformers remain, but class transformers change between transformer and gyrator.

of the same physical class and only change the effort/flow ratio, the impedance (sec. 2.2).

Figure 8.22 is a "fantasy graph", that is a graph without definite conventions for explaining the difference of through and across variables. There are 2 junctions at the top and on the bottom (connected electric wires or p-junctions) and 3 connections or links in between, each with an element represented by a circle. Connections are called links or arcs in the literature of linear graphs. Here we find two kinds of variables:

1. The variables that need 2 contact points for measuring, in electricity the voltage and in hydraulics the pressure: across variables.
2. The variables that require to cut a connection or link for measuring, electrically the current and hydraulically the flow: through variables.

The main differences of both classes of variables seems not to be the measuring process but the question of origin (zero point) and absolute values:

- For an across variable the origin has no significance, only the difference of the two values or meter readings.
- For a through variable there exists a definite zero. If it is reached, the connection can be cut without disturbing the system.

In mechanics the speed (rotation frequency) is an across variable and force (torque) a through variable. Consequently, with the force flow analogy, all efforts

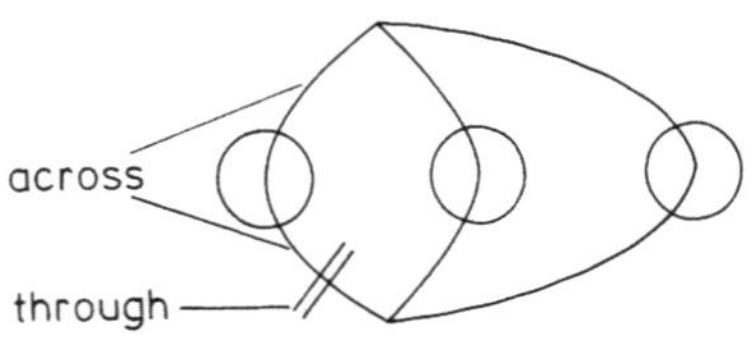

Figure 8.22 A "fantasy" graph with 3 connections with elements represented by circles. Across variables are measured applying a meter across an element, through variables require cutting a link (voltmeter and ammeter in the electric case).

(including speed, rotation frequency) are across variables and all flows (including force, torque) are through variables.

Linear graphs are built on the distinction of through and across variables and we explain them by the simple example of Fig. 8.23. At the left there is a parallel resonance circuit with I-, C-, and R-elements written systematically, that is without grounding the lower p-junction. Abstracting the elements and the s-junctions we obtain the simpler representation at the right which is a linear graph. It carries less information and is more compact than a Bondgraph. The power signs of Bondgraphs become the full arrows in the links, also called arcs. Further, the linear graph can be divided into two subgraphs, the tree and co-tree. This is equivalent to assigning causalities which become integer if all C-elements are in co-trees. The junctions of the fantasy graph of Fig. 8.21 are called nodes in the literature on linear graphs.

Linear graphs are built around mathematical graph theory which includes the notion of trees and co-trees: they assist to write computer code. So, very powerful programs exist, especially for complex systems.

For a Bondgrapher the general way to write a linear graph is to:

- Write a systematic Bondgraph in the force flow analogy without grounding a node.
- Replace all s-junctions by an arc, removing the connected one-port elements.
- Assign power directions (full arrows in linear graphs) and choose causalities by selecting a co-tree: The linear is graph ready.

An example with an electric motor drive appears in Fig. 8.24 as a circuit at the top and as a Bondgraph at the bottom in force flow analogy. The electric side is as usual, but the motor is now a transformer and the flywheel a C-element. Figure 8.25 shows the passage to a linear graph by the abstraction of Fig. 8.23 or to a compact Bondgraph by grounding the lower notes. As seen, a linear graph has no way to represent a two-port (although some authors use broken lines) so the graph divides into disconnected subgraphs. Linear graphs are more compact as already noted: jokingly they have been described as Bondgraphs (in force flow analogy) almost starved to death.

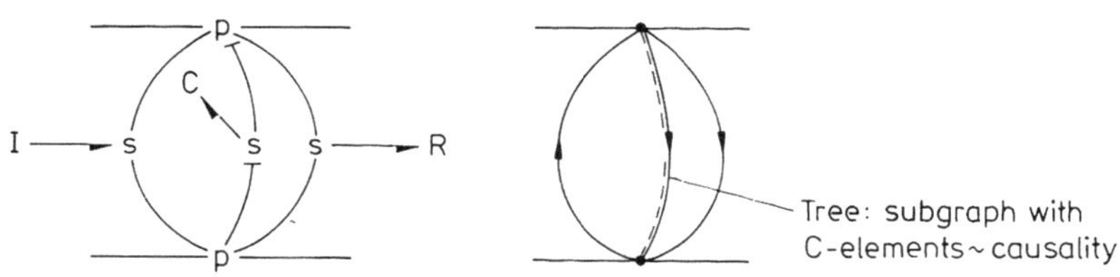

Figure 8.23 Left a small systematic Bondgraph, right the passage to the linear graph by abstraction of the elements and s-junctions. The broken lines indicate that the link is part of a tree (a subgraph). Selecting a tree is equivalent to selecting causality in a Bondgraph.

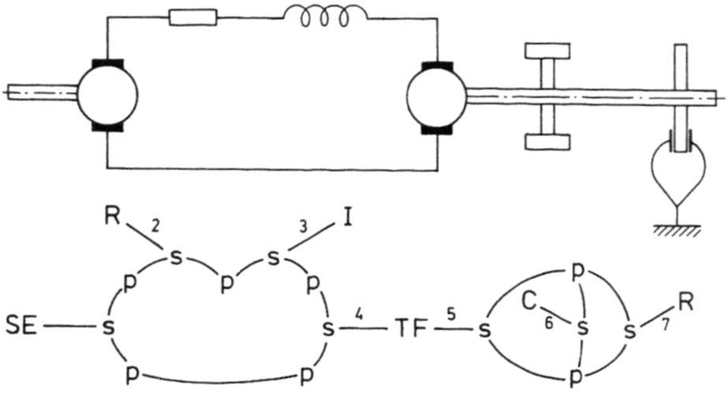

Figure 8.24 Example of an electric motor drive with a Bondgraph in force flow analogy where the electric motor becomes a transformer and the inertia a C-element.

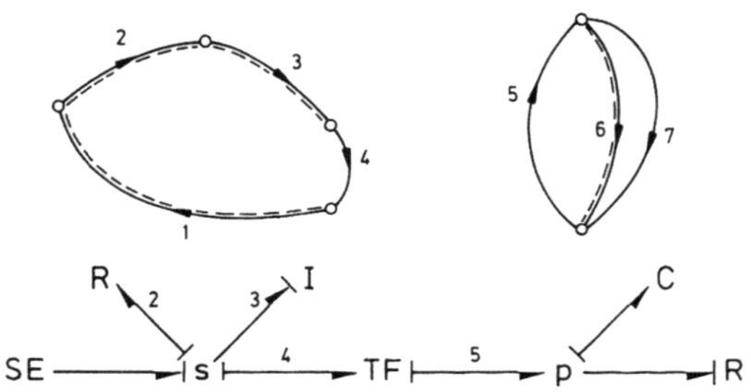

Figure 8.25 Passage from Fig. 8.24 to the linear graph by abstracting elements and s-junctions and to the Bondgraph by grounding the lower notes. After this the graphs are quite dissimilar.

Although the author definitely prefers Bondgraphs as a more intuitive and practical representation, many publications and programs exist for linear graphs. Consequently, common study and cross fertilization can also help for Bondgraph simulation. For some research of their relation consult Birkett, Roe, Thoma 1988; it is based on the combinatorial structure of the junctions and uses the mathematical concept of a matroid, common to both graphs.

The author has also seen Bondgraph simulations in the force flow analogy and mathematically there is no preference between both. We should even note that the systematic writing in force flow analogy is simpler because all points of definite efforts are given a p-junction with elements connected behind an s-junction. The special procedure for mechanics, where all points of definite velocity receive an s-junction is not necessary (compare sec. 3.3).

8.4.1 Portality Table

For added light on analogies, Bondgraph elements can be arranged in the matrix or table form of Fig. 8.26 with their energy or power handling arranged vertically and the number of ports horizontally. We distinguish here between power conservation, where power fluxes sum up to zero at each instant, and energy

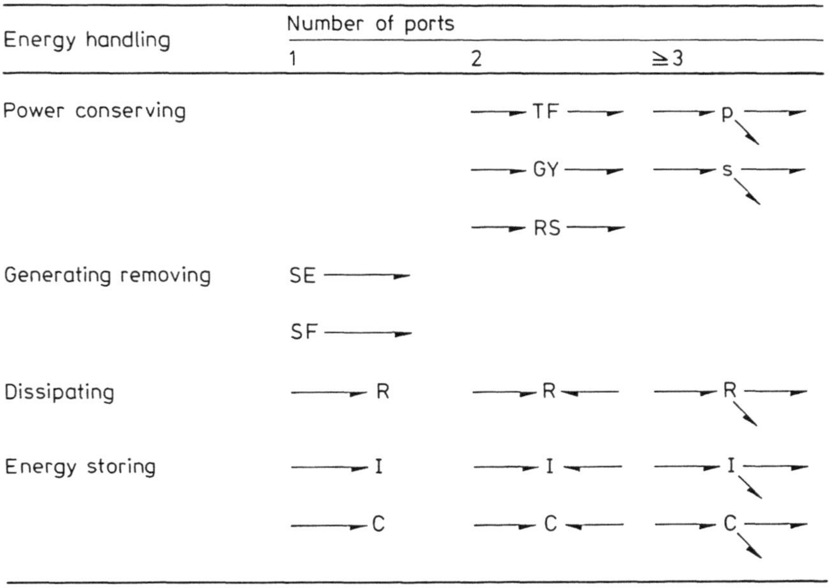

Figure 8.26 Portality table with fields for Bondgraph elements, with power/energy handling vertical and number of ports horizontal. Passing between the analogies the elements change but stay within their fields.

conservation, where power can be absorbed but will change the state of the element and needs to be released to return to the original state. For power conservation two or more ports are needed.

Some observations to the portality table of Fig. 8.26:

- Sources generate power from the outside or remove it (if reverse driven).
- Dissipating elements turn power into the thermal domain and are irreversible. For multiport R-fields, only overall dissipation is irreversible, but power can come out from a non-thermal bond if more power enters through another non-thermal bond (Fig. 2.35).
- Including thermal power, the dissipating elements (R-elements and R-fields) become power conserving.

Returning to the different analogies, elements change as we have seen but always stay in the same field of the portality table. So power and energy handling and the number of ports remain invariant under a change of analogy.

One might speculate what elements could be placed inside the empty fields of Fig. 8.26, but this is remote from practical Bondgraph simulation. As an historical note, the portality table is based on a night discussion with R.C. Rosenberg who also coined the word as an expression of the Paynterian spirit. This is the spirit underlaying all Bondgraph reticulation combining physics, computation and intuition.

8.5 The Problem of Alternating Current Bondgraphs

Near the end of this book we want to open the study of electric motors for alternating currents, enlarging the purely mechanical simulations of sec. 3.4 and of Fig. 5.7. From electric machine theory (Say 1976) we know that the 3-phase current in an electric machine produces a rotating field in the stator (similar to Fig. 5.4 for DC machines) which tends to drag the rotor along; this is the basic idea behind the Bondgraph 3.15 of the asynchro motor. There it induces a voltage in the resistance circuit of the rotor and the current produces the torque. Voltage and electrical frequency are proportional to the slip, that is the difference of electric frequency in the stator and the mechanical (rotation) frequency of the shaft.

If the rotor has its own magnetic moment as in synchro machines, it is coupled by a mechanical C-element (Fig. 5.7). In both machine types we have abstracted from the number of pole pairs (f.i. two pairs with half the rotation frequency and double the torque): they can be represented by a transformer.

Mechanical torque induces a real (in phase) voltage (usually called EMF) in the stator, as necessary for power transfer. Simulation of the starting or normal run of electric machines is of great interest, including influence of electric resistance, magnetic stray fluxes and mechanical load. Electric textbooks use equivalent circuits for electric quantities, but are quite primitive for interconnection to mechanical elements. The equations, however, are well known and reasonably simple.

The problem is connected with computation: We want to calculate the envelopes of time dependent alternating currents and voltages, not the instantaneous quantities. For the envelope, electric engineering has developed the notion of average (RMS) current and voltage and the graphical method called phasor diagram. A key point is that the power in a pair of AC wires depends on the phase angle between current and voltage in addition to their amplitudes. Hence in a Bondgraph the bonds need to carry this information as well.

Figure 8.27 shows a basic reticulation, where the AC bonds with voltage, current and phase appear in two different notations:

- above as parallel continuous and broken lines
- below a continuous line with a sign across like the Greek letter φ, the symbol for electric phase angle.

Since parallel lines are harder to draw than signs on single lines (bonds) the author prefers the φ symbol.

The main element is the frequency transformer which connects AC electric power to mechanical rotary power. The rotation frequency is rigidly equal to the electric signal frequency. The mechanical torque produces the in-phase part of the stator current drawn from the voltage source at the left, in addition to the magnetization current known from transformers (Fig. 5.12). So we have a rigid rotation frequency, rigid as long as the electric frequency of the supply remains

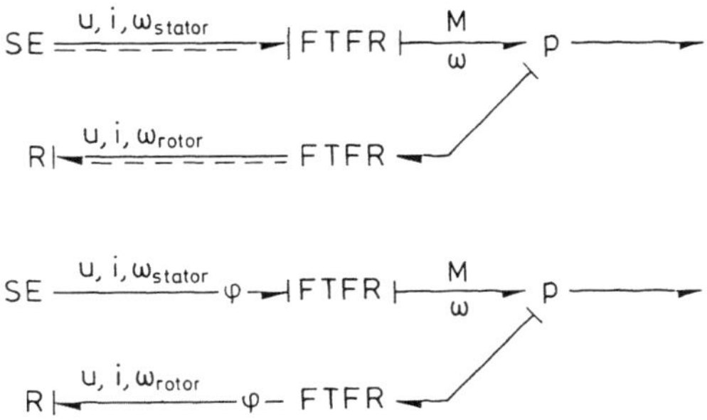

Figure 8.27 Reticulation of an electric AC machine with frequency transformers FTFR that produce mechanical rotation frequency equal to electric frequency. In an asynchro machine another FTFR appears from the p-junction and leads to the R-element at the left. The electric AC bonds also contain phase information and are represented by a continuous and a broken line on top or a Greek letter φ symbol on the bond (bottom). In case of a synchro machine, the R-element would need to be replaced by a C-element for the fixed magnetization of the rotor.

constant; further a conductance proportional to the torque leading to the indicated causalities of Fig. 8.27. Nevertheless, the FTFR is power conserving.

Figure 8.27 contains a further FTFR below, coupled mechanically through a p-junction. It also has a rigid coupling of frequency and feeds an AC voltage to the windings of the rotor with the electric frequency equal to the mechanical slip frequency. The rotor resistance and inductance determine the electric current which corresponds to the mechanical torque. The voltage is proportional to the slip frequency. So the FTFR element has some of the properties of a gyrator.

The figure displays the basic relations, but should be further developed. Details of figures and symbols might depend on the computer program for envelope calculation, which is outside the capability of present TUTSIM. In the view of the authors, all relations in electric machine theory (including tearing of matrices of Gabriel Kron, DQ transformation of Park) just cry out for a Bondgraph representation. As reported by Dean Karnopp, Henry Paynter spoke about the penetrating but Olympian methods of Gabriel Kron.

So a desirable representation should replace the complex matrix equations by clear, impressive and beautiful Bondgraph symbols that also help to program the computer. Perhaps some reader will be inspired.

8.5.1 Epilogue and Reflection

With the challenge of AC-Bondgraphs we reach the end of the main part of our book. There follow some worked examples to illustrate special techniques and appendices with auxiliary information.

Now the reader will appreciate the Bondgraph way of life of Fig. 3.1: We combine technical knowledge, computational techniques and the analogies between the disciplines. Further, we always study power flows and thermodynamic reversibility in addition to the interaction of signals from control engineering. The power and beauty of the Bondgraph way comes from the harmony between these well integrated approaches.

9 Further Worked Examples

9.1 Hydraulic Examples

9.1.1 Water Pump

Figures 9.1a to 9.1d show a single acting one cylinder water pump; the Bondgraph includes the dynamics of the intake valve for investigation of erosion of its seat in a form similar to Fig. 4.23. The delivery valve is represented (without dynamics) as nonlinear resistor R35 (R351 and FNC35). Realistic parameters are used.

The parameters refer to a very small pump with 4 mm piston diameter 6 mm stroke running with 50 Hz. It delivers against a pressure of 0.2 MPa from SE42. The valves are balls with 2 mm diameter and maxi opening of 0.4 mm.

Figure 9.1d gives one period of 20 msec with sinusoidal piston flow Q11, cylinder pressure p13 and valve position x231. Interesting for the original research are the oscillations of the valve position when hitting the maxi opening stop during intake and on reseating on delivery. The oscillations are controlled by increasing R22, in practice by changing the form of the valve body.

On reseating the valve deforms the seat by about 20 micrometers due to the finite stiffness of the seat. This stiffness controls with the valve mass I24 the time step to about 8 microseconds and a similar step is required from C13 with R25 when open. So for run up 10 microseconds were used, and 5 microseconds for final presentation.

9.1.2 Jetpipe Servovalve

Control valves in hydrostatic engineering are often of the spool type (Fig. 4.9). The spool is moved by a magnetic actuator with a hydraulic preamplifier; it consists of a movable jet with two collector holes that direct fluid to one or the other end faces of the spool. Figure 9.2a shows the principle and Fig. 9.2b enlarges one collector hole. The valve is miniature: jet diameter 0.2 mm, holes diameter 0.3 mm, spool diameter 7 mm and length 50 mm.

The pressure/flow characteristic of the collector holes is determined from a momentum balance and leads to the equation on Fig. 9.2b below (Thoma 1988). The pressure has a term independent of flow minus a term porportional to flow,

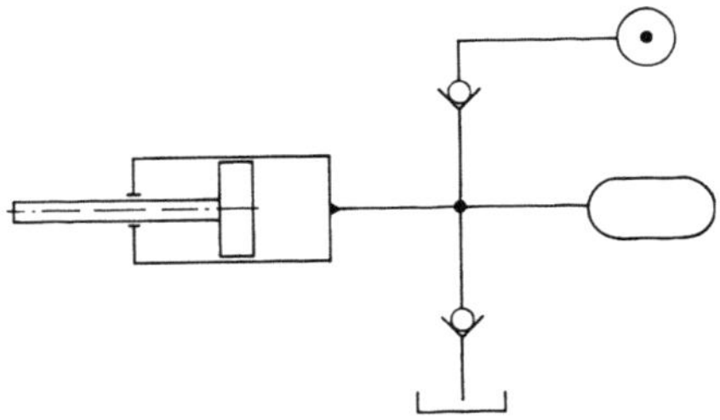

Figure 9.1a Schematic of a simple water pump with check valves.

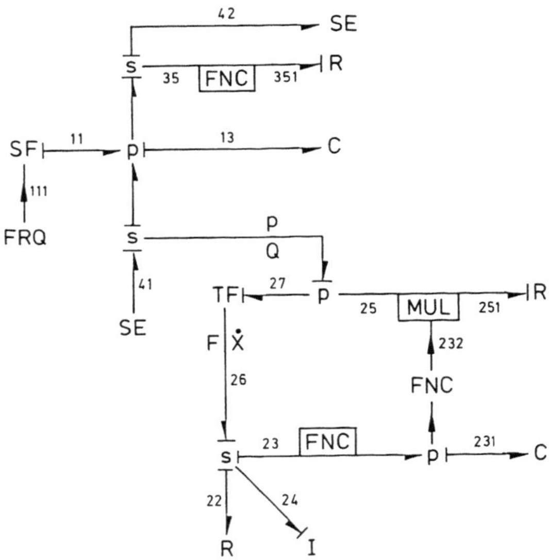

Figure 9.1b Bondgraph of the water pump. The upper (delivery) check valve appears as nonlinear resistor, but the lower (intake) valve includes the dynamics of the valve body. Hence there is a nonlinear spring (C23) and a modulated resistor (R25).

both functions of the area α of the injet (relative to the hole area). This corresponds to an effort source with a resistor in parallel, namely SE31 and R32 on one face, SE41 and R42 on the other; both work in push-pull to move the main spool.

Here the procedure is the inverse of the usual: having the equations we construct the Bondgraph of Fig. 9.2c, collect the parameters in the paracheck 9.2d and write the TUTSIM listing 9.2e. Simulation results are on Fig. 9.2f, above for a step current input i18, below for an impulse current. The magnetic actuator is in the upper part of Fig. 9.2c. The spool action in the lower part includes the modulated R62 for leakage and damping; it is proportional to the ideal flow Q61 or spool displacement x531. All hydraulic resistors are linearized for pressure but nonlinear by their modulations.

```
Timing:    5.000E-06   ,DELTA  ;    0.0210000   ,RANGE
PlotBlocks and Scales:
Format:
     BlockNo,  Plot-MINimum,   Plot-MAXimum;  Comment
Horz:      0 ,   0.0000      ,   0.0200000  ; Time
  Y1:     11 , -20.0000      ,  20.0000     ; peak piston flow
  Y2:    231 ,-500.000E-06   , 500.000E-06  ; valve seat
  Y3:     13 ,   0.0000      ,   0.2500000  ; cyl capacity
  Y4:        ,               ,              ;

 -11.7000           11 SF      111                    ;peak piston flow
   0.0110000        13 C        11       25       27  ;cyl capacity
                               -35
   0.1593950
   0.0400000        22 R        24                    ;valve friction
                    23 FNC     231
     1 -400.000E-06    -10.0000
     2    0.0000         0.0000
     3  400.000E-06      0.0100000
     4  800.000E-06     10.0000
  36.000E-06        24 I        26      -23      -22  ;valve mass
   0.0175500
                    25 MUL     251      232           ;real conductance
   2.5400           26 TF      -13       41           ;valve area
   2.5400           27 TF       24                    ;idem
                    35 FNC     351
     1   -1.0000         0.0000
     2    0.0000         0.0000
     3    1.0000       380.0000
   0.0000           41 CON                            ;intake pressure
   0.2000000        42 SE      421                    ;upstream pressure
  50.0000          111 FRQ                            ;drive 50 Hz
   0.0000
   1.0000          231 C        24                    ;valve seat
 -16.200E-06
                   232 FNC     231
     1   -2.0000         0.0000
     2    0.0000         0.0000
     3    2.0000         2.0000
 950.000E+03       251 R       -13       41           ;valve conductance
   1.0000          351 R        13      -42           ;delivery valve
   1.0000          421 CON                            ;file bonsi91a.sim
```

Figure 9.1c TUTSIM listing for the water pump.

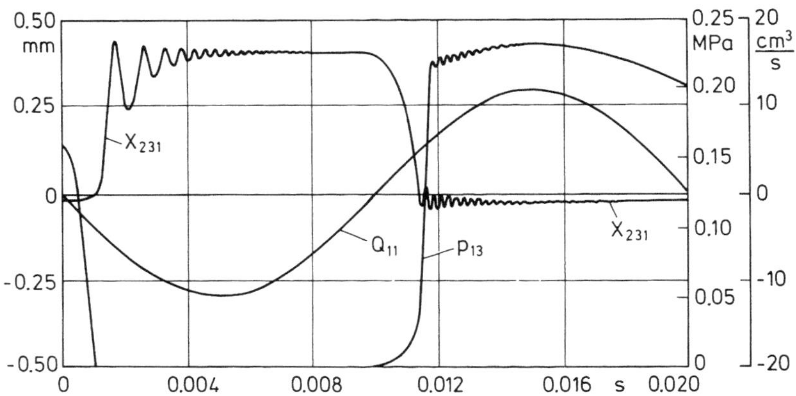

Figure 9.1d Simulation of the water pump with visible oscillations of the body of the intake valve.

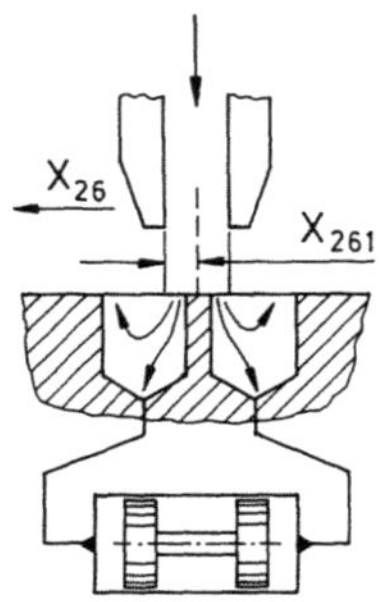

Figure 9.2a Schematic cross-section of the jet pipe and the main spool (in smaller scale).

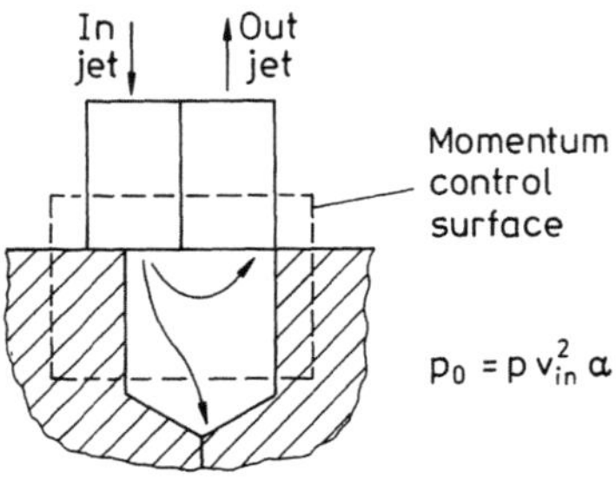

$$p_0 = \rho v_{in}^2 \alpha (1 + \alpha (1 - \alpha)) - 4 \rho v_{in}^2 \frac{Q_0}{A_0} \alpha (1 - \alpha / 2)$$

Figure 9.2b Enlarged collector holes with the control surface for momentum balance.

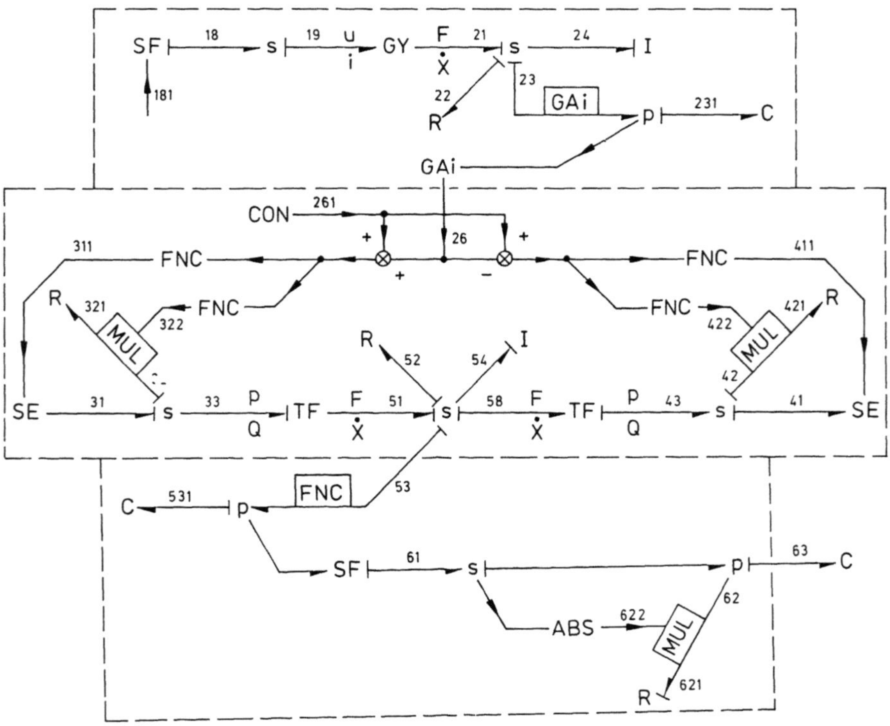

```
Timing:    5.000E-06  ,DELTA  ;    0.0210000   ,RANGE
PlotBlocks and Scales:
Format:
     BlockNo,  Plot-MINimum,  Plot-MAXimum;  Comment
Horz:      0 ,   0.0000      ,   0.0200000  ; Time
  Y1:    329 ,  -4.0000      ,  16.0000     ; left face pressure
  Y2:    231 ,   0.0000      , 100.000E-06  ;
  Y3:    531 ,  -0.0010000   ,   0.0010000  ; Spool posit
  Y4:    429 ,  -4.0000      ,  16.0000     ; right face pressure


   0.4000000        18 SF      181                    ;Electric current source
   0.0800000        21 GY       18                    ;Actuating force
   0.1000000        22 R        24                    ;Jet pipe friction
 400.0000           23 GAI     231                    ;Restraining spring
 500.000E-06        24 I        21      -22     -23
   0.0000
   1.0000           26 GAI     231
  42.0000           31 SE      311                    ;Left Pressure source
                    32 MUL     321      322           ;Series resist fr mom balance
  38.5000           33 TF       54                    ;Spool end face
  42.0000           41 SE      411                    ;right right press source
                    42 MUL     421      422           ;right spool end face
  38.5000           43 TF       54                    ;Right spool end face
  38.5000           51 TF       31      -32
  71.0000           52 R        54                    ;Spool friction
                    53 FNC     531
     1   -0.0014000      -9.200E+03
     2 -400.000E-06     -92.0000
     3  400.000E-06      92.0000
     4    0.0014000       9.200E+03
   0.1000000        54 I        51      -52     -53   ;Spool mass (bad consci)
                               -58
   0.0000
  38.5000           58 TF       41       42
   2.440E+06        61 SF      531
                    62 MUL     621      622           ;Modu para resist = 2*psp/Q61
   1.0000           63 C        61      -62           ;Load model, provisio
   0.0000
   0.0000          181 PLS                            ;file bonsi91b.sim
   0.2000000
   0.5000000
   1.0000          231 C        24
   0.0000
  60.000E-06       261 CON                            ;opening displac at rest posit
                   311 FNC      26      261
     1 -200.000E-06       0.0000
     2    0.0000          0.0000
     3   25.000E-06       0.0048810
     4   50.000E-06       0.0522810
     5   75.000E-06       0.1747940
     6  100.000E-06       0.3212330
     7  125.000E-06       0.4341870
     8  150.000E-06       0.5049530
     9  175.000E-06       0.5424600
    10  200.000E-06       0.5440000
    11  300.000E-06       0.5440000
  10.8000          321 R        33                    ;left basic resist
                   322 FNC      26      261
```

◁

Figure 9.2c Bondgraph of the jet pipe servovalve. The symmetry comes from the push-pull action of the collector holes and spool end faces.

Bond paracheck (jetpipe JEP 2. SiM)

Element			Value	Unit	Parametre output	valcode
SF 18	exciting current	0.4	0.4	A	i_{18}	
GY 19, 21	magnetic actuator		0.08	N/A	F_{21}	
R 22	jet pipe friction		0.1	$\frac{N}{ms}$	F_{22}	
C 23 GAi 23	restraining spring		2.5 E-03	$\frac{m}{N}$	C_{23} GAi_{23}	0.0025 400.0
I 24	jetpipe mass 0.5g		500 E-06	$\frac{NS}{m/s}$	$\dot{X}_{24}$	
SE 31 41	source part = 4 p_{supply} for s 311 = 1		42.0	MPa		
R 321 421	resistace part for s 322 = 1		10.8	$\frac{MPa}{\mu LA}$	p_{321}	
FNC 311	area modulation (source)		from separate BASIC programme			
FNC 322	area modulation (resistance)					
TF 33, 51 43, 58	spool end faces dia = 7 mm		38.5	mm^2	F_{51} Q_{33}	38.5
R 52	spool friction		71.0	$\frac{N}{m/s}$	F_{52}	71.0
I 54	spool mass		15 E-03	$\frac{Ns^2}{m/s}$	$\dot{X}_{54}$	
FNC 53	stop force of spool	-1.4 -9200.0	-0.4 -92	mm N	0.4 92.0	1.4 9200.0
R 621	parallel resistor with Q 62 = 1		42.0	$\frac{MPa}{\mu LA}$	Q_{621}	0.023
CON 261	opening displacement at rest		60.0	µm	X_{261}	60.0 E-06

Figure 9.2d Parameter checklist for the jet pipe servovalve.

```
        1 -200.000E-06      0.0000
        2    0.0000         0.0000
        3   25.000E-06      0.0048460
        4   50.000E-06      0.0486680
        5   75.000E-06      0.1426410
        6  100.000E-06      0.2324530
        7  125.000E-06      0.2909700
        8  150.000E-06      0.3240510
        9  175.000E-06      0.3406250
       10  200.000E-06      0.3450000
       11  300.000E-06      0.3450000
                    329 SUM     31     -32          ;left face pressure
311.0000            411 FND    -26     261
 10.8000            421 R       43                  ;right basic resist
322.0000            422 FND    -26     261
                    429 SUM     41      42          ;right face pressure
  1.0000            531 C       54                  ;Spool posit
  0.0000
  0.0231000         621 R       63                  ;Unit parallel resist
                    622 ABS     61                  ;Sign correction f modu
```

Figure 9.2e Listing for the simulation.

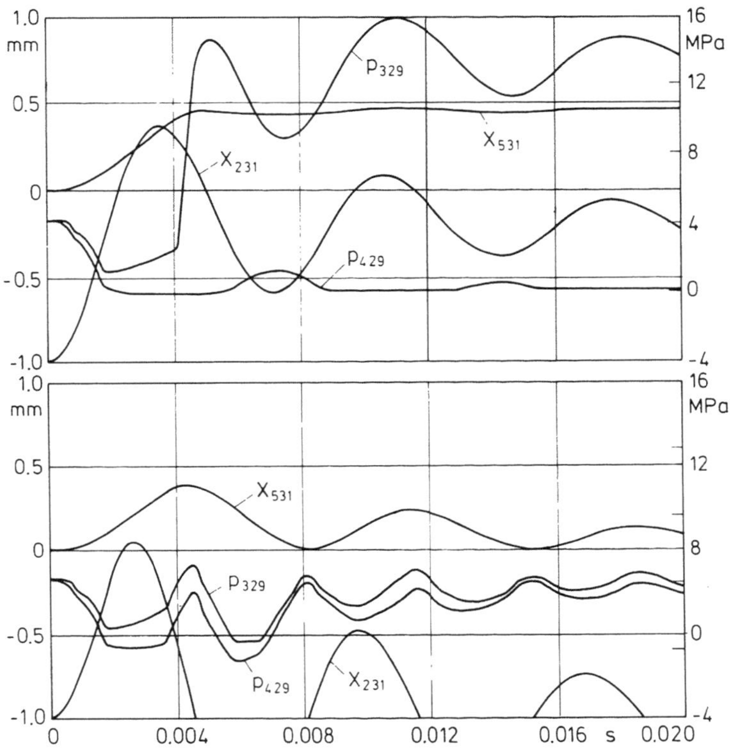

Figure 9.2f Simulation result at the top with step excitation, below with a pulse of 2 msec duration.

9.1.3 Pressure Regulator Valve for Breathing

The pressure regulator valve for Figs. 9.3a and b is placed between a high pressure supply and a load. A typical application is the breathing apparatus for divers under water (supplyequipment); here the sources are the compressed air bottles and the load the lung of the diver. Similar valve systems are used also in industrial hydraulics (fluid power).

The pressure at the load is maintained by the piston with the connected slide valve. A spring pushes it upwards, opening the valve. The load pressure pushes downwards: if it rises the valve tends to close. Regulating systems of this kind have a certain importance in qualitative simulation (Bobrow 1985) and are used as example in Thoma (1989). Their nonlinear effects are essential since qualitative simulation focusses on three states: closed, floating and fully open.

Figures 9.3a is the schematic and 9.3b the Bondgraph. We find again, similarly to Fig. 4.19, the modulated resistor in conductance causality and the nonlinear C23 for the spring and the stops at the ends of the stroke. For convenience in

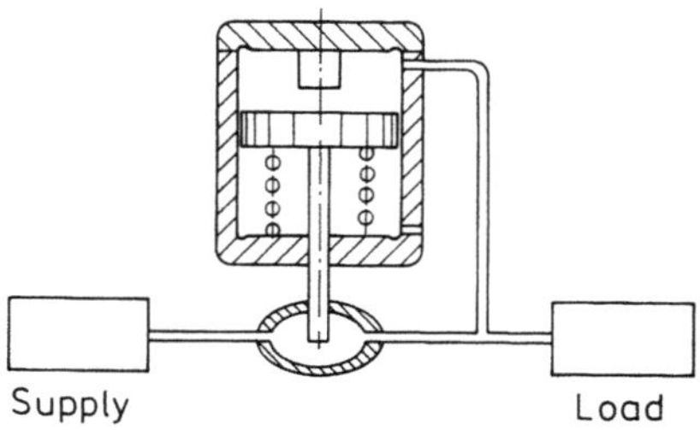

Figure 9.3a Schematic cross-section of the pressure regulator valve between high pressure supply and load.

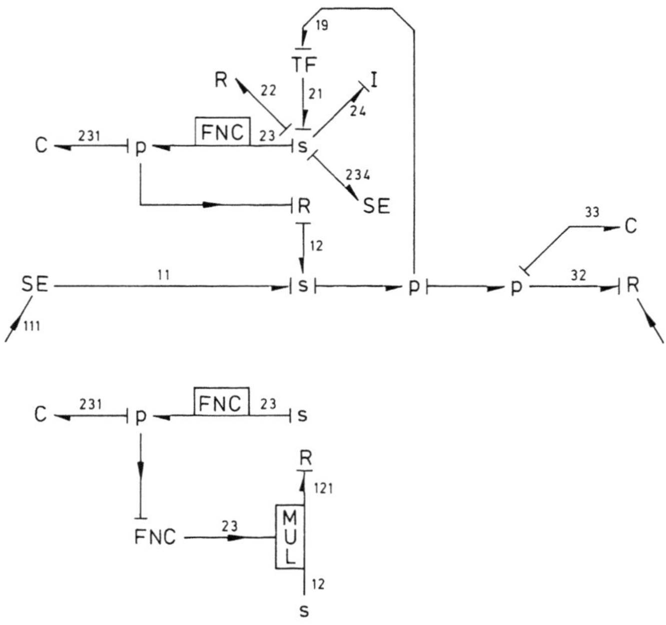

Figure 9.3b Bondgraph of the pressure regulator system, with details below.

adjusting parameters, SE234 represents the preload of the spring. Details appear on Fig. 9.3b below.

Figure 9.3c contains at the top the simulation result and below the listing. We find violent transients on start. After about 0.5 seconds the flow demand is much reduced by giving a low conductance to R32. After the new transients, Q12 settles much lower, but p33 remains almost constant, as expected from a regulator.

Let us note in conclusion that the behavior depends not only on the regulator valve, but on supply and load as well, in short on the entire pressure system.

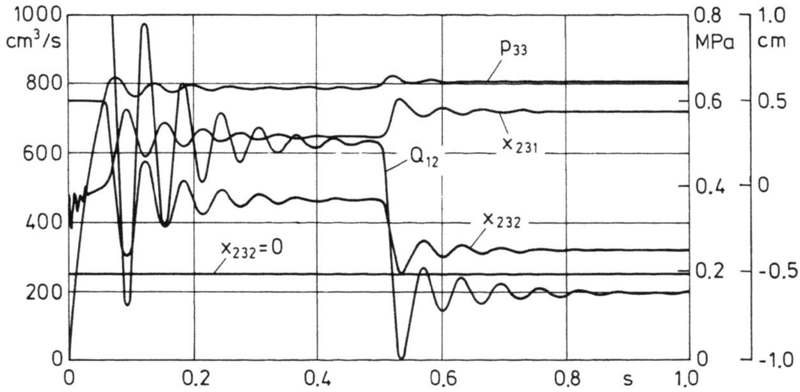


```
Timing: 500.000E-06   ,DELTA  ;    1.0100      ,RANGE
PlotBlocks and Scales:
Format:
     BlockNo,  Plot-MINimum,   Plot-MAXimum;  Comment
Horz:      0 ,   0.0000      ,   1.0000     ; Time
  Y1:     12 ,   0.0000      ,   1.000E+03  ; throttle resistance + flow
  Y2:     33 ,   0.0000      ,   0.8000000  ; capacity of load
  Y3:    231 ,  -1.0000      ,   1.0000     ; spring as integrator
  Y4:    232 ,  -0.2500000   ,   0.7500000  ;

   1.0000           11 SE     111                      ;supply pressure
                    12 MUL    121     232              ;throttle resistance + flow
   3.1400           19 TF      24                      ;piston 20 dia
   3.1400           21 TF      33                      ;piston 20 dia
   0.0050000        22 R       24                      ;piston friction
                    23 FNC    231
     1   -0.5000000      -6.2800
     2    0.0000          0.0000
     3    0.5000000       0.1570000
     4    1.0000          6.2800
  30.000E-06        24 I       21     -22      -23    ;piston mass + speed
                             -234
   0.0000
   1.000E+03        32 R       33                      ;outflow resist o load
 200.0000           33 C       12     -19      -32    ;capacity of load
   0.0000
   1.0000          111 CoN
   8.000E+03       111 R       11     -33
   1.0000          231 C       24                      ;spring as integrator
   0.0000
                   232 FNC    231
     1   -0.5000000       0.5000000
     2    0.0000          0.5000000
     3    0.5000000       0.0000
     4    1.0000          0.0000
   1.8800          234 CON                             ;spring preload
```

Figure 9.3c Simulation result and listing of the pressure regulator. After 0.5 sec load demand decreases substantially; this leads to new transients but pressure remains almost constant.

9.2 Thermofluid Examples

9.2.1 Hydropneumatic Fatigue Testing Machine

We return here to the hydropneumatic accu of Fig. 2.4 as part of a fatigue testing machine. There is a hydrostatic machine on Fig. 9.4a with periodically variable displacement (about 2 secs period) which sends and removes fluid into the accu and a ram that compresses a test piece. So the installation is a generator of alternating force and much more compact and powerful than the cylinder of the water pump in sec. 9.1

The question is whether the periodic compression of the gas generates any steady entropy (heat) currents in the irreversible R-elements of the walls of the accu. This would be in addition to the periodic temperature change of the gas which is a thermal/hydrostatic C-field. What needs to be avoided is dangerously high temperature. A pseudo Bondgraph with temperature and heat flow is insufficient and we need a real Bondgraph with entropy currents similar to Fig. 7.6. As a reminder, entropy current is heat flow divided by absolute temperature.

However, Fig. 7.6 is valid only if entropy flow and temperature differences never reverse signs, but in the test machine they do so. Then the new entropy equals dissipated power divided by the lower temperature; is injected in the capacity with the lower temperature. Hence we need the Bondgraph in Fig. 9.4b.

The wall of the casing is reticulated by the segment method with two capacitors, C13 and C23 and a resistor between. It is denoted RESO (REsistor/SOurce) and shown in detail in Fig. 9.4b below. Here we have the following effects for this entropy resistor:

- as in Fig. 7.6, RESO generates entropy flow, equal to the dissipated power (temperature difference times incoming entropy flow) divided by the lower of

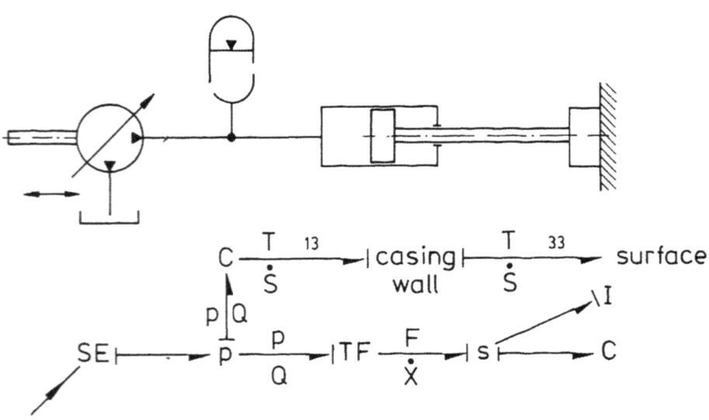

Figure 9.4a Schematic circuit and general Bondgraph of a hydropneumatic fatique test system. The word "casing wall" should be replaced by TUTSIM blocks for entropy conduction and generation.

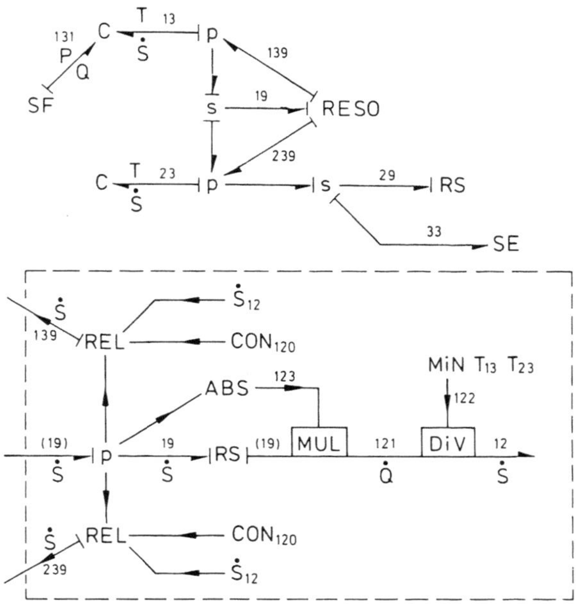

Figure 9.4b Detail Bondgraph with entropy current for investigating entropy generation (heating) with repeated cycles.

```
Timing:    0.0500000  ,DELTA  ;   20.1000      ,RANGE
PlotBlocks and Scales:
Format:
     BlockNo,  Plot-MINimum,   Plot-MAXimum;  Comment
Horz:      0 ,    0.0000      ,  20.0000      ; Time
  Y1:     13 , 280.0000       , 320.0000      ; upper capaci
  Y2:     23 , 280.0000       , 320.0000      ; lower capaci
  Y3:    239 ,  -0.5000000    ,   0.5000000   ; lower S switch
  Y4:    139 ,  -0.5000000    ,   0.5000000   ; upper S switch

                    12 DIV    121      122           ;new S = entropy flow
 10.0000            13 C      -19      131     139   ;upper capaci
300.0000
 20.0000            19 R       13      -23           ;Ṡ conductance
 10.0000            23 C       19      -29     239   ;lower capaci
300.0000
 10.0000            29 R       23      -33           ; Ṡ conduc t enviro
290.0000            33 SE     331
  0.0000           120 CON
                   121 MUL     19      123           ;new Q = heat flow
                   122 MIN     13       23           ;calcul lower temp
                   123 SUM     13      -23           ;calcul temp dif
100.0000           131 SF     132                    ;squeezing volu flow
  0.5000000        132 FRQ                           ;sinusoidal pressure
  0.0000
  0.0000           139 REL    120      120      12   ;upper S switch
                              123
  0.0000           239 REL     12       12     120   ;lower S switch
                              123
  1.0000           331 CON                           ;file bonsi92.sim
```

Figure 9.4c Listing with entropy current: it is much more complex than the pseudo Bondgraphs of Figs 7.7 and 7.8.

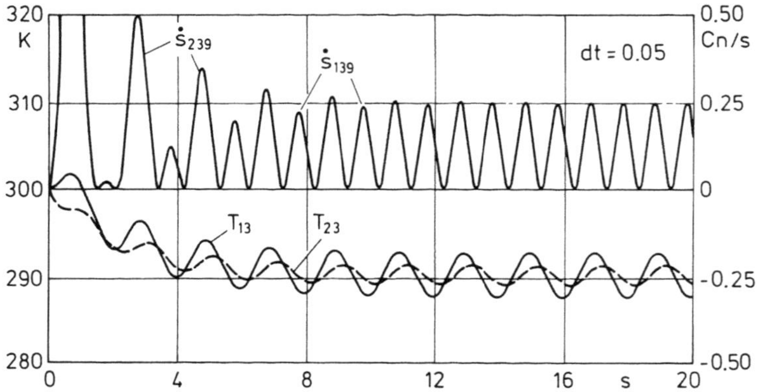

Figure 9.4d Simulation with entropy current.

the temperatures T13 and T23; this is calculated by the MIN122 and the DIV12 blocks.

- The new entropy flow is thrown into the capacity with lower temperatures by the relays REL139 and REL239.

Figure 9.4c shows the corresponding TUTSIM code. The gas as two-port C-field is approximated by C13 and SF131 for the sequeezed out entropy on periodic excitation from FRQ132. Note also that temperatures are always positive and need to be expressed in the Kelvin scale (absolute values). This is a point where entropy flow differs essentially from electric current and a starting point for advanced research. Otherwise the parameters are simplified.

Figure 9.4d contains the simulation result: the temperatures go down from the initial 300 K nearer to the 290 K of the environment. So no steady entropy flow seems to be generated. Prolongating the simulation, such a flow becomes visible, but is not large nor dangerous.

The entropy resistor RESO is connected not only to bond 19 but also to bonds 139 and 239. This indicates that entropy flow depends on T13 and T23 singly, not only on their difference. There is a curious analogy with chemical reactions (Fig. 7.17) far from equilibrium, where the molar flow depends also on the chemical potentials separately, not only on the difference (Thoma–Atlan 1977). In both cases, the condition for equilibrium or no flow is the equality of efforts, that is temperatures or chemical potentials, respectively.

The above thermofluid system is needed for the accu problem, but complex: generally the pseudo simulation of Fig. 7.7 is preferable. Entropy is a great fundamental variable of physics but not very suitable for calculation of conduction.

A1 Organization of Bondgraph Courses

Bondgraphs are a universally applicable system representation and there is much latitude in organizing courses. As a modelling tool they should come before control engineering, although Bondgraphs are really a concentrated block-diagram, fundamental to controls.

The author has experience in 3 different courses, called Modebond, Simbond and Controlint, respectively.

Modebond or modern machine design by Bondgraphs is a one term, a 33 hours graduate course, ME 744 at the University of Waterloo, Canada. This book was written as a textbook for it and is based on the informal lecture notes, regularly updated from 1978 to 1987, together with the 1975 book of one of us.

Simbond is a 2 days, 16 hours intensive course for engineers in industry, developed originally in France, but taught in Switzerland, Germany, USA and Canada and Japan (2 days in spring 1989). There is one computer for each two of the gentils participants (maximum 24) and the course is run from special lecture notes written for people with industrial experience. The book will be used as additional material.

Controlint is a standard introduction to controls taught by the author as in ME 360 at the University of Waterloo for third year undergraduate students. It starts the normal way with blockdiagrams, Laplace Transform etc. After about 2 weeks students are impressed by the difficulty of writing blockdiagrams. Then Bondgraphs are taught for about 2 weeks on special lecture notes. The young students understand essential points like causality quickly and show better examination results than with conventional matter like Nyquist and Bode plots and stability. This book is recommended for further reading.

The graduate Modebond Course comprises 4 assignments and computer exercises with a final project. The assignments are:

1. Write a Bondgraph for a simple installation proposed by the teacher.
2. Write a Bondgraph for a system from your experience or interest.
3. Write a Bondgraph for a problem given by the teacher, include a paracheck and realistic parameters and simulate it on a computer.
4. Make a project with complete simulation, realistic parameters and paracheck from your area of interest or experience; it can be a continuation of assignment 2.

Assignments 2 and 4 stimulate the creativity of the students and bring many interesting questions to the teacher, who acts as a consultant. Only very rarely the teacher needs to give a project. The contrary often happens. The project of the student is so complex that it cannot be finished in the available time (ca 20 hours); it needs to be cut down.

A2 Note on Bondgraph Standards and Graphics

Bondgraphs are a graphical systems representation that helps intuitive understanding and computer programing. Many people worldwide work with Bondgraphs and tend, not surprisingly, to develop their own notations for the above goals. The drawback of personal or institutional notations is that they become harder to understand for people from different communities. The analogy with the dialects of human languages is striking.

In the author's view, the primary standards are contained in the books of Karnopp-Rosenberg (1975 and 1968) and any deviation should be made only most reluctantly and against tangible advantages, just as a young lady should give up her virtuous principles only for the same reasons.

Thoma Standards

We use in this book the following deviations from and complements to the Karnopp–Rosenberg (1975) standards:

JUNCTIONS: We use "s" and "p" for series and parallel junctions instead of the "1" and "0" notation. Having taught courses with both standards, and consulted, we finds ours vastly superior as mnemonic aid and for cleaner computer code. It is for programs like ENPORT whilst TUTSIM is indifferent in this respect. So people remember the similarity with series and parallel circuits or arrangements. Nobody seems to worry that this is not really so in mechanics.

BLOCK-BONDCRAPHS: They were published in Thoma (1985) originally as an aid to write efficient computer code for nonlinear components. They turned out to be graphically impressive, even beautiful in many nonlinear systems. Indeed it seems that most interesting system are nonlinear. Manual linearization is time consuming and error prone. Refer also to a most interesting paper of Karnopp (1975) on power in linearized physical systems.

MULTIBONDS: We deviate from the Breedveld (1985) standard in graphical details: for a multibond we put a ring around the line instead of using two parallel lines. This because rings are easier to draw, important in complex Bondgraphs. Also we show the splitting into single strand (Fig. 8.1) a little differently for more graphical appeal. The reader is invited to compare the interesting paper of Breedveld with this book.

Recommendation to Other Bondgraphers

To repeat, deviations from standards should be made most reluctantly, but can be unavoidable. In particular, there may be a need to define symbols for special elements. Here we urge to use four or more letters, since standard elements have one to three letters. So the reader sees immediately that it is a special element from another Bondgrapher.

Further, four letters carry more information and make remembering, indeed understanding the publication easier. In a way they are the transition to a word Bondgraph.

We have observed the above recommendation in this book with the words HEXA for HEateXchAnger in sec. 7.2 and FTRF for Frequency TRansFormer in 8.5. We could have left out the last letter in both cases, but against our urgent recommendation.

A3 Table of the Main Bondgraph Elements

Bonds and Connections

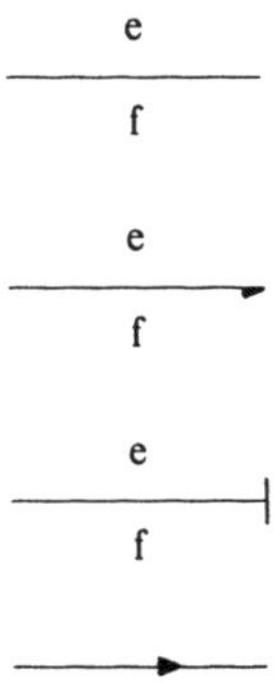

Simple bond	
Power flux $\dot{E}$	
causality effort pushes to right, flow points to left	
Connection one variable e or f neglected	

Elements with One Bond (One-ports)

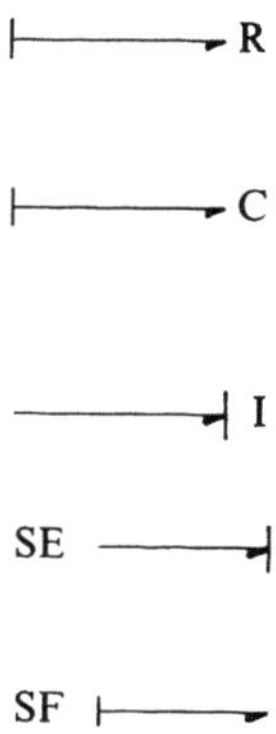

	Equations
R-element aborbs or dissipates $\dot{E}$ (power)	$e = R \times f$ $f = e/R$
C-element stores E (energy)	$e = \frac{1}{C}\int dt \times f$ $f = C\, de/dt$
I-element stores E	$f = \frac{1}{I}\int dt \times e$ $e = I\, df/dt$
effort source supplies $\dot{E}c$	$e = e(t)$
flow source supplies $\dot{E}$	$f = f(t)$

Note: Causality of sources is mandatory otherwise short-circuit!

Elements with Two Bonds (Two-ports)

		Equations
1 —⊣TF—⊣ 2	Transformer (modulated)	$e_2 = re_1$
—⊣MTF—⊣ (modulated by r)	conserves $\dot{E}$	$f_1 = rf_2$
1 ⊢—GY—⊣ 2	Gyrator (modulated)	$e_2 = rf_1$
⊢MGY—⊣ (modulated)	conserves $\dot{E}$	$e_1 = rf_2$

Elements with Three Bonds (Three-ports)

—⊣P—⊣ (third bond)	Parallel junction, denoted also —P—	$e_1 = e_2 = e_3$
	conserves $\dot{E}$ denoted also	$f_1 + f_2 + f_3 = 0$
1 ⊢—S⊢— 2, 3	Series junction, denoted also —S—	$e_1 + e_2 + e_3 = 0$
	conserves $\dot{E}$ denoted also	$f_1 = f_2 = f_3$

Advanced Elements

⊢—C—⊣	C-field stores E	$e_1 = f_{01}(q_1, q_2)$ $e_2 = f_{02}(q_1, q_2)$
—⊣I⊢—	I-field stores E	$f_1 = f_1(P_1, P_2)$ $f_2 = f_{02}(P_1, P_2)$
—R—	R-field dissipates $\dot{E}$	$e_1 = f_{01}(f_1, f_2)$ $e_2 = f_{02}(f_1, f_2)$
e, f —RS⊢— T, $\dot{S}$	RS-field conserves $\dot{E}$, generates entropy irreversible	$\dot{S} = ef/T$

Glossary of Frequent Terms in English, French and German

English	French	German
Note: Terms in brackets are not recommended, but used sometimes.		
actuator	actionneur	Betätigung
algebraic loop	boucle algèbrique	algebraische Schleife
angular momentum	impulsion angulaire	Drehimpuls
armature current	courant d'armature	Ankerstrom
bearing	palier	Lager
block diagram	schéma blocs	Signalflussplan
Block-bondgraph	Block-bondgraphe	Block-Bondgraph
bond	bond, liaison	Bindung
bondgraph	bondgraphe	Bondgraph
bondgraph	(graphe à liaison)	(Bonddiagram)
C-element	element-C	C-element
C-field	champ-C	C-feld
calculating step	pas de calcul	Rechenschritt
capacitor	condensateur	Kondensator
causality	causalité	Kausalitaet
clutch	embrayage	Kupplung
conductance	conductance	Leitwert
connection	connéxion	Verbindung
damper, Lanchester	amortisseur Lanchester	Torsionsdaempfer
derivative causality	causalité dérivative	derivative Kausalitaet
desirable causality	causalité désirable	wünschenswerte Kausal
displacement	déplacement	Verschiebung
drag efficiency	rendement de trainèe	Reibungs-Wirkungsgrad
efficiency	rendement	Wirkungsgrad
effort source	source d'effort	Spannungsquelle
Eigenvalue	valeur propre	Eigenwert
energy	energie	Energie
entropy	entropie	Entropie
equivalent circuit	circuit equivalent	Ersatzschaltbild
excitation	excitation	Erregung
feed back	retroaction	Rueckfuehrung
flow source	source de flux	Flussquelle
flywheel	volant	Schwungrad
frame, Eulerian	repère, Eulerien	Bezugssystem, Eulerisches
friction	frottement	Reibung
gyrator	gyrateur	Gyrator
hydrostatic machine	machine hydrostatique	hydrostatische Maschine
I-element	élément-I	I-element
I-field	champ-I	I-feld

English	French	German
impedance	impèdance	Impedanz
integral causality	causalite integrale	integrale Kausalitaet
jet pipe	tuyeau a jet	Strahlrohr
junction	jonction	Verknüpfung
linear graph	graphe lineaire	Netzwerk-Graph
mandatory causality	causalité necessaire	notwendige Kausalitaet
modestack	pile de modes	Mode-Haufen
modulated transformer	transformateur modulé	modulierter Transformator
momentum	impulsion	Impuls
multiport	multiport	Mehrport
network	réseau	Netzwerk
one-port	monoport	Einport
one-port	(Element à une porte)	(Eintor)
parallel junction	jonction parallele	Parallel-Verknüpfung
Planetary gear	engrenage planetaire	Planeten-Getriebe
power	puissance	Leistung
power direction	direction de puissance	Leistungs-Richtung
R-field	champ-R	R-feld
ram pump	bélier (hydraulique)	Widder (hydraulisch)
relief valve	clapet de surpression	Druckventil
resistance	resistance	Widerstand
series junction	jonction série	Serien-Verknüpfung
slip efficiency	rendement de glissement	Slip-Wirkungsgrad
slipper	Patin	Gleitschuh
slipping (parameters)	ramener	ueberziehen
state space	espacè d'etat	Zustandsraum
torque	couple	Drehmoment
transformer	transformateur	Transformator
transmission	transmission	Getriebe
tree, Bondgraph	arbre de bondgraphes	Bondgraph-Baum
two-port	diport	Zweiport
two-port	(Element a deux portes)	(Zweitor)

References

ACSL, 1988, a simulation program for VAX and IBM PC from Mitchell and Gauthier Assoc., Concord, MA USA 01 742 or Rapid Data Ltd, Arundel, West Sussex, UK BN18 9DJ.

Atlan, H., Katchalsky, A., 1973, Tellegens Theorem for Bondgraphs and its Relevance to Chemical Networks, Currents of Modern Biology, Vol 5, pp 55–65.

Breedveld, P.C., 1985, Multi Bondgraph Elements in Physical Systems Theory, Journ. Franklin Institue, Vol 319, pp 1–36.

Bobrow, D.G. (editor), 1985, Qualitative Reasoning about Physical Systems, MIT Press.

CAMP 1988, written by Dr. Jose Granda, California State University, Sacramento, is available from the same sources as ACSL.

ENPORT, 1988, from Rosencode, 3900 Capital City Boul. #102, Lansing, MI, USA, 48 906. "Written originally by Prof. Rosenberg".

Falk, G., Ruppel, W., 1976, Energie und Entropie, Springer Verlag "most interesting, unfortunately not in English, but German.

FANSIM, 1988, Program for Fourier and Modal Analysis by Ken Fyfe, from TUTSIM Products, 200 California AV, Palo Alto, CA USA 94 306.

Fleischmann, R., 1980, Einfuehrung in die Physik, 2 ed, Physik Verlag, Weinheim, Germany. Original publication Zeitschrift für Physik, vol 129, pp 377–400, 1951 and vol 138, pp 301–308, 1954.

Fyfe, K., Thoma, J. U., Ismail, F., 1986, Structural Identification from time Domain Simulations, 4th IMAC conference, Los Angeles, pp 745–750.

Gear, C.W., 1971, Numerical Problems in Ordinary Differential Equations, Prentice Hall.

Hogan, N., 1985, Impedance Control: An Approach to Manipulation, Jour. Dyn. Systems and Control, ASME, Vol 107, pp 17–24.

Karnopp, D.C. 1977, Power and Energy relations in Linearized Physical Systems, Journ Franklin Inst., Vol 303, pp 85–98.

Karnopp, D.C., 1979, Pseudo Bondgraphs for Compressible Thermofluid Systems, Trans ASME, JDSMC Vol 101 pp 201–204.

Karnopp, D.C., Rosenberg, R.C., 1975, System Dynamics: A Unified Approach, Wiley.

Karnopp, D.C., Rosenberg, R.C., 1968, Analysis and Simulation of Multiport Systems, MIT Press.

Margolis, D.L. 1985, Bondgraph Modelling of Lumped and Distributed Systems, Journ. Franklin Institute, Vol 319 pp 134.

Martin, H.R., McCloy D., 1980, the Control of Fluid Power, 2nd ed. Ellis Horwood/John Wiley.

MATLAB, 1988, Program for matrix manipulation on a PC by Mathworks Inc, 158 Woodland Str. Sherborn, MA USA 01 770.

Say, M.G., 1976, Alternating Current Machines, Pitman.

Schmidt, G., 1980, Simulationstechnik, Oldenburg (in German).

Sueur, C., Dauphin-Tanguy, G., 1990, Structural Controllability and Observability by Bondgraphs, Journ. Franklin Institute, Special number on Bondgraphs, spring 1990.

Takahashi, Y., Rabins, M., Auslander, D., 1970, Control Addison Wesley.

Thoma, J.U., 1970 A, Modern Oilhydraulic Engineering, Trade and Technical Press, Morden, Surrey, England.

Thoma, J.U., 1970 B, Bondgraphs for Thermal Energy Transport and Entropy Flow, Journ. Franklin Institute, Vol 292, pp 109–120.

Thoma, J.U., 1975, Introduction to Bondgraphs and their Applications, Pergamon Press, Oxford, England.

Thoma, J.U., 1985, Block Bondgraphs for Nonlinear Systems, Trans. Soc. Compu. Simulation, Vol 2, pp 249–255.
Thoma, J.U., 1988, Fluid Mechanics, Bondgraphs and Jetpipe Servovalves, Proc. 12th IMACS World Congress, Paris. publication expected from North-Holland.
Thoma, J.U., 1990, Bondgraphs for Qualitative and Quantitative Systems Modelling, in P. Fishwick, P. Lucker, editors, Qualitative Simulation Modelling and Analysis, series Advances in Simulation, Springer Verlag, 1990.
Thoma, J.U. Atlan, H., 1977, Network Thermodynamics with Entropy Stripping, Journ. Franklin Inst. Vol. 303, pp 319–328.
Thoma, J.U. Atlan, H., 1985, Osmosis and Hydraulics by Network Thermodynamics and Bondgraphs, Journ. Franklin Inst, Vol. 319, pp 217–226.
Thoma, J.U., Richter, D.B., 1987, Simulation of Fluid Pipes with Modal and Segmented Methods, TRans. Soc. Compu. Simulation, Vol 3, pp 337–349
Tiernego, M.J., Bos, A.M., 1985, Modelling Mechanical Systems with Multi Bondgraphs, Journ. Franklin Institute, Vol 319, pp 37–55.
TUTSIM, 1988, program developed by V. Meerman, from Meerman Automation, POB 154, NL 7160 AC Neede, Netherlands, or from TUTSIM Products, 200 California AV, Palo Alto, CA USA 94 306.

Subject Index

The numbers refer not to pages, but to the figures, where the subject can be found nearby. An asterisk* indicates a definition or a specially interesting entry.

A-Matrix Box 6.1
Accumulutator, Hydropneumatic 2.4, 4.6
Across Variable 8.21
ACSL 6.1
Algebraic Loop 6.17
Analogies 8.20, 8.21
Antenna 1.2
B-Field, Magnetic 5.3, 8.16
Battery 7.21
Bearing, Hydrostatic
Slipper hydrostatic
Blackbody Radiation 7.15
Blockdiagram 1.4, 1.7
Blocked Shaft 3.19
Bondgraph 1.4, 2.1
C-element 2.4
C-Field 2.30, 2.32
CAMP 6.1
Capacitor 2.4
Carnot Efficiency 2.4
Carnot Engine 7.2
Carrousel of Variables 2.2, 2.6
Causal Conflict 4.11, 6.15
Causality 2.13
Chemical Reaction 7.17, 7.19
Clutch 3.13, 3.13*
Compact Units 6.25
Conscience, Bad 6.17
Convection 7.11, 7.14
Convection 7.14
Coupled Flow 7.20
Desirable Causality 2.13
Diamond Equality 2.36, 2.37
Diamond Theorem 2.37, 2.38
Disk Brake 2.3, 4.16
Drawings 1.1
Dynamometer 3.7
Dynamometer, Electric 3.8
Efficiency 3.18
Efficiency of Hydrost Machin 4.14
Eigenvalue Box 6.1
Electric Motor, AC 3.15, 8.27
Electric Motor, DC 5.3
Enport 6.1
Enthalpy 7.11
Entropy 7.1
Entropy Generation 7.4
Eulerian Frame 7.2, 7.10
FANSY, FANSIM 6.1
Fluid Pipe 8.6, 8.10
Flywheel 2.5, 8.24
Force/Effort Analogy 2.1, 8.20
Free Shaft 3.19
Frequency Transformer 8.27
Gravity Force 2.7
Gyrator 2.10
H-Field, Magnetic 8.16
Hammer 8.13, 8.14
Heat Exchanger 7.10
HEXA Heat Exchanger
Hydrostatic Machine 4.14, 4.15
Hydrostatic Transmission 4.13
Hysteresis 8.16*, 8.18
I-Element 2.5
I-Field 2.34
Impact 8.16
Inductance, Hydraulic 4.8
Inductor 2.5
Irreversible, Thermodynamic 2.3, 7.4
Jetpipe Valve 9.2
Junction, Parallel 2.11
Junction, Series 2.11
Karnopps Theorem 8.4
Lagrangian Frame 7.9
Lanchester Damper 3.11
Linkage 6.19, 8.2*
Loudspeaker 5.10
Magnetic Actuator 5.11
Mandatory Causality 2.13
Matroid 8.24
Micro Enport 6.1
Modal Method 8.7*, 8.10
Mode Stack 8.8, 8.10
Multibond Transformers 8.4
Multibond 7.20, 8.1
Networks 1.5
Node 8.21
One-Port 2.3
Onsager Symmetry 2.8
Operating Limit 1.3

Osmosis 7.17
Parallel Excited Motor 5.4
Parameter Determination 3.19
Parameter Slipping 6.9, 6.10
Parasites 3.18
Paynterian Spirit 8.26
Phantasy Graph 8.21
Phase, in AC 8.27
Planetary Gear Reducer 3.9
Portality 8.25, 8.26
Pseudo Bondgraph 2.7, Table 2.2
Quasi Power 8.5
R-Element 2.3
R-Field 2.35
Radar Antenna 1.2
Railway Wagons 3.5
Ram Pump 8.10
Ram, Hydraulic 2.23, 4.3*, 4.9
Relief Valve 4.9*, 4.10
Relief Valve, Hydraulic 4.10
reversible (thermodynamic) 2.8, 7.2
Segment Method 8.6
Series Excited Motor 5.4
Servomotor, Hydrostatic 4.9
Shock tube 8.10
Shock Waves 8.11
Slip 3.12
Slipper, Hydrostatic 4.21, 4.23
Slipping, Parameters 6.9, 6.11
Solar Energy 7.13
Source, Effort 2.7, 2.8
Source, Flow 2.7, 2.8
Spool Valve 4.9, 9.2
Steady State 6.4*, 6.6
Strands (in Bonds) 8.1
Tellegens Theorem 8.5
Through Variable 8.21
Transformer 2.9
Transformer, Real 5.13
Transmission 3.17
Transmission, Hydrostatic 4.14
Tree, Bondgraph 6.27
Tree, Linear Graph 8.23
Turbine 7.11
Tutsim 6.1
Two-Port 2.9
Voltage Generator 1.6, 2.7
Water Pump 9.16
Word Bondgraph 2.1

MIX
Papier aus verantwortungsvollen Quellen
Paper from responsible sources
FSC® C105338

If you have any concerns about our products,
you can contact us on
ProductSafety@springernature.com

In case Publisher is established outside the EU,
the EU authorized representative is:
Springer Nature Customer Service Center GmbH
Europaplatz 3, 69115 Heidelberg, Germany

Printed by Libri Plureos GmbH
in Hamburg, Germany